Retaining W
Engineering

Diversity drives better business results; however, despite decades of effort, women make up only 15% of engineers. *Retaining Women in Engineering: The Empowerment of Lean Development* approaches the problem of women leaving engineering from a systems-level perspective to change the way engineering is done and level the playing field between men and women.

This book utilizes the six principles of Lean Development and draws from the learnings of the field of medicine, recognizing that access to a vast amount of written knowledge is an important part of a physician's learning process. Using these principles, the book provides leaders with concrete strategies and methods to change the way engineering is done and learning occurs. Integrated within the book are "gray box stories" which describe two different worlds that engineers work in: that of traditional development and that of Lean Development. These stories underscore the ways that the gender confidence gap, bias, and stereotypes affect a female engineer's career. Additionally, the book highlights how the methods of Lean Development strengthen an individual's ability to control their learning and career, and a leader's ability to coach others more effectively. Ultimately, this results in more capable teams. Furthermore, not unlike the marine chronometer (a clock) which solved the centuries old challenge of establishing the longitudinal location for a ship at sea, this book finds the "clock" that levels the playing field between men and women.

This book will help leaders at every level within an engineering firm, as well as women engineers and managers who want to grow to their full potential, and others who care about gender equity.

Women of STEM: Innovation and Leadership Series

Series Editor: Pamela McCauley

The aim of this new series is to become a leading and premiere resource to published STEM, Innovation and Leadership books that address current and emerging topics by female authors. The focus is on timely STEM topics and innovations associated with the UN Sustainable Development Goals and Engineering's Grand Challenges make this a series with STEM expertise and "heart." This is because the problems identified in these two broad categories are about using STEM knowledge to positively impact the global society. Thus, books focused in these areas have the potential to teach relevant, timely and needed solutions to national and global problems. The presence of a female focused series has the potential to reduce the intimidation that many female authors experience at the idea of publishing a book. The fact that the series is "female focused" is essentially an "invitation" for women to publish their books and a validation that there is an interest and need to disseminate their achievements in the form of a book.

The focus on STEM will allow for a broad array of book titles under this series. The emphasis on Engineering's Grand Challenges and the UN SDG's will allow women working in virtually any STEM discipline to understand how their research/technical projects/STEM knowledge can be applied to these key areas of global importance.

Retaining Women in Engineering
The Empowerment of Lean Development
Robert N. Stavig and Alissa R. Stavig, MD

Retaining Women in Engineering

The Empowerment of Lean Development

Robert N. Stavig
Alissa R. Stavig, MD

Foreword by Durward K. Sobek II

CRC Press is an imprint of the
Taylor & Francis Group, an informa business

First edition published 2023
by CRC Press
6000 Broken Sound Parkway NW, Suite 300, Boca Raton, FL 33487-2742

and by CRC Press
4 Park Square, Milton Park, Abingdon, Oxon, OX14 4RN

CRC Press is an imprint of Taylor & Francis Group, LLC

The opinions expressed in this book reflect the personal opinions of the authors and contributors and do not reflect the opinions of our employers. None of the information in this book should be taken as medical advice or take the place of mental health treatment.

The stories of interactions and experiences between unnamed individuals throughout the book are invented and although represent real world situations are all fictional. Any similarity to individuals or groups of people and their experiences or behaviors are accidental.

Library of Congress Cataloging-in-Publication Data

Names: Stavig, Robert N., author. | Stavig, Alissa R., author.
Title: Retaining women in engineering : the empowerment of lean development / Robert N. Stavig and Alissa R. Stavig.
Description: First edition. | Boca Raton, FL : CRC Press, 2023. | Series: Women of stem: innovation and leadership series | Includes bibliographical references and index.
Identifiers: LCCN 2022019227 (print) | LCCN 2022019228 (ebook) | ISBN 9781032071886 (hbk) | ISBN 9781032071855 (pbk) | ISBN 9781003205814 (ebk)
Subjects: LCSH: Women in engineering.
Classification: LCC TA157.5 .S73 2023 (print) | LCC TA157.5 (ebook) | DDC 331.4/8620--dc23/eng/20220912
LC record available at https://lccn.loc.gov/2022019227
LC ebook record available at https://lccn.loc.gov/2022019228

ISBN: 978-1-032-07188-6 (hbk)
ISBN: 978-1-032-07185-5 (pbk)
ISBN: 978-1-003-20581-4 (ebk)

DOI: 10.1201/9781003205814

Typeset in Times
by Deanta Global Publishing Services, Chennai, India

To Landon and Grace

Contents

Appendices

To the Reader

The premise of this book is that the low retention rate of women in engineering is driven by the way the work is done and a lack of independent access to a firm's knowledge (meaning, dependency on their peers to share their knowledge). Progress to solve these issues will therefore require changing the way engineers learn and the work methods they use.

Based on our research, this is the first book or publication that directly links the low retention rate of women engineers to the methods of doing the work and the methods of learning. Current research and publications examining the low retention rate of women engineers have focused on areas such as support from mentors, workplace bias, addressing bias in promotion and hiring processes, gender confidence, and providing career growth. Many of these efforts target changing the behaviors of individuals rather than changing the system of engineering. Thus far, these efforts have not provided the desired results. This book promotes the need for a change to the engineering system.

In order to understand and then address the low retention rate of women engineers, we will focus on four key areas:

1. Describing the challenges women face in engineering from a holistic perspective
2. Comparing the training and learning environment of engineers and physicians (women now make up more than 50% of medical school attendees)
3. Comparing traditional development and lean development
4. Identifying concrete strategies and tactics that a firm can make in order to change the engineering system through the principles of lean development

This book is written for three populations:

1. Leaders at every level who are responsible for developing and retaining women engineers while simultaneously delivering on business objectives
2. Women engineers or managers who want to grow into their full potential
3. Individuals who care about gender equity and are searching for ways to reach it

We hope that the methods proposed in this book, to change the system of engineering, enable every woman to make engineering a lifelong career – just as they expected when they received their degrees. Through the leadership of management and the efforts of individual male and female engineers, we believe this can be accomplished.

Bob Stavig, Brush Prairie, WA
Alissa Stavig MD, Billings, MT

Foreword

Why did you decide to pick up and open this book? Perhaps it's because your company has implemented a slew of programs to improve gender equity in the workplace, yet women are still far outnumbered among the engineering ranks, and you want to understand why and are possibly looking for a new approach. Perhaps you have recently stepped into a leadership role within your product development organization and are looking for guidance to retain the incredible talent you currently have. Perhaps you are tired of seeing highly capable female colleagues leaving your organization for other opportunities and you are trying to understand why and what you might be able to do about it. Or maybe you are one of the many women in engineering who are questioning their career choices, wondering if anyone understands your reasons or if leaving engineering is your best option.

If you fit one of these categories, or even if not but are just looking for fresh insight into the topic, then you've picked the right book.

I've been an engineering faculty member at a land-grant university for over two decades. Throughout that time, recruiting young women into engineering has been the priority. One key reason is that, according to research, more diverse teams and workplaces tend to be more innovative and outperform their less diverse counterparts. The same is true for classrooms. Another reason is that not attracting women excludes the talent of half the population. A third reason is simply that we should be striving to make engineering and the places where engineers work more equitable. However, while there have been pockets of success – some engineering majors have achieved near-parity between men and women, there are significantly more female faculty, K-12 outreach programs have exploded, and others – we still have a long way to go. Engineering, as a broad discipline, is still far from reflecting the demographics of the general population, at least within the United States.

In reviewing this manuscript, I learned that the uphill recruitment challenge is made significantly steeper when the engineering workforce lacks gender diversity. When young women search for rewarding careers, the lack of role models and the profusion of male-dominated workplaces would give many pauses: If so few women are working in engineering, maybe there are good reasons for that, so maybe engineering isn't for me. A key strategy in recruiting women into engineering is providing them with a vision of a rewarding career with examples of people like them who are enjoying meaningful, rewarding work.

Coauthors Bob Stavig and Dr. Alissa Stavig are focused on this end of the pipeline; making it more attractive for women to continue with engineering as a career. Bob is a retired engineering manager with many years of experience in one of the world's premier engineering companies that have international reputation for innovation and quality. Alissa is a practicing medical doctor well on her way to a productive, successful career. This father–daughter team brings a unique string of insights sparked by the stark differences in how medicine and engineering cultivate early-career professionals in their respective fields.

The authors acknowledge that the myriad programs which progressive companies have adopted to improve gender equity are well and good, indeed necessary, but insufficient. To retain women in engineering-related career paths, the authors argue that companies need to go further. They need to change how the work gets done. Engineering workplaces, they point out, remain somewhat impervious to bias literacy training and other efforts to promote gender equality because the work methods used in engineering result in structural inequities for women. Though women are the focus of this work, these work methods also contribute to structural inequities for other underrepresented groups. The following story illustrates how work methods institute structural inequities.

A few years ago, my daughter called home frustrated by how a group project was going. At the time, she was a first-year student in an engineering program. She was assigned to work in a three-person group with two young men, also engineering majors. She was frustrated because, in her view, they would not listen to her ideas. Based on my experience with young male engineering students and my understanding of knowledge-based product development, I suggested she try a different strategy. Rather than trying to convince her team members of the merit of her ideas, I suggested she try responding with, "That's an interesting idea. Let's try it!" Then figure out quick ways to test the ideas and see how they work. A while later, I asked her how that group project was going. "A lot better," she said. She tried that approach, and they had fun breaking things. But after several rounds of this, they grew tired of looking stupid and agreed to try some of her ideas. Through it all, they learned a lot and arrived at a successful solution that earned a decent grade, and she felt she had made valuable contributions to the group's efforts.

How does this illustrate structural inequities in engineering work? The normal method of selecting design ideas in these team settings is through discussion and debate, driven by opinions, experience, and personalities. As you'll learn in this book, such a method gives men an advantage. Compared to women, men tend to be overconfident, more assertive, and have experiences that, on the surface at least, appear better aligned with engineering problems. (My daughter, for example, was more inclined to curl up with a good book in her free time than play with Legos.) When the team switched to a knowledge-based approach, i.e., using experiments to vet ideas, it created a more level playing field where the young woman's ideas could be explored on par with the others. In other words, it created a more equitable team environment.

This story is a microcosm of the kinds of work methods at play in many engineering workplaces. In much greater depth using careful causal analysis, the authors argue that the key to retaining women in engineering is improving job satisfaction, and to do that companies need to change their work methods, become more knowledge-driven, and encourage role models. They further argue that several of the marque approaches from lean product and process development (LPPD, or Lean Development for short) provide the type of work methods and knowledge-driven practices which can produce that more level playing field.

I met Bob some 15 years ago when his division brought me in to learn about Lean Development, which has been a subject of my research for many years. Bob and his team applied, learned from, and built upon that work with success and mastery ever since that time (without much guidance from me, I might add!) until his recent retirement. His vast experience forms the basis for the realistic and vivid scenarios used to illustrate points throughout the book. Alissa's contributions are equally seminal as the contrasts between medicine and engineering turned the key and opened the door to a whole new way of thinking about engineering work and professional development of engineers.

The value of this book is not just in explaining how certain workplace practices subtly and unwittingly lead to marginalization and dissatisfaction of women in engineering workplaces. The authors also present concrete tools and practices that can be implemented for meaningful change and point to additional resources for those who would like to dig even deeper. For anyone interested in this important topic, this book will be worth your time.

Durward K. Sobek II, PhD
Bozeman, MT

Acknowledgments

This book would not have been possible without the 20-plus years of research conducted and published in articles and books about the challenges women face in STEM occupations. This research provided a foundation for developing a deeper understanding of the issues faced by women in STEM. Two books of particular value to our conceptualization of these issues were Virginia Valian's *Why So Slow? The Advancement of Women* (1998) [1] and Joyce K. Fletcher's *Disappearing Acts: Gender, Power, and Relational Practice at Work* (1999) [2]. We also want to acknowledge the early researchers and authors in the area of Lean Development who provided us with the principles of Lean Development which allowed us to propose a system-based change which we believe is required to better enable the success of women in engineering.

In regard to actually writing the book and the research process, this book would not have happened without the support, contributions, and encouragement of Deb Stavig. Deb was critically involved throughout the book's development as a sounding board, brainstormer, researcher, editor, and overall fierce supporter. We would like to thank Jessica Wanke both for her contribution to Chapter 3 and for her hours spent editing. Her brutal editing of Bob's initial writing attempts and teaching him how to approach writing a basic paragraph helped move the book forward. We are very appreciative of Deb Blakewood and Rose Elley for their time and dedication in writing Chapter 12, bringing in their personal experiences and insights as well as the experiences of many other women in engineering. We would like to thank Durward Sobek II for his review of the book and the insights he provided as well as his early research in Lean Development, which was critical in the formulation of our proposed system-based change, and his writing of the forward. We had several readers who provided valuable feedback and comments either on chapters or the entire book. Specifically, we would like to thank Jane Gagliardi, Rathnayaka Mudiyanselage Gunasingha, Venkata Jonnalagadda, Hope Sparks, Kyle Wanke, Nate Wilson, Paul Wilson, and Catherine S. Wolff.

In addition to thanking the early researchers and authors in the area of Lean Development, Bob would specifically like to thank Michael Kennedy for his published book in 2003 (Bob's first exposure to Lean Development in 2003), the late Allen Ward (whom Bob was able to sit next to for an entire day during a 2004 workshop on Lean Development), and Durward Sobek II. These three individuals changed the course of his engineering career. Bob would also like to thank the dozens of engineers who have been supportive of the use of Lean Development while asking thought-provoking questions about the methods. Specifically, he would like to thank Kynan Church, Lynn Collie, Catie Gould, Toni Hugill, Ryan Khor, Christie Larson, Huy Le, Kristi Lo, Mark Miranda, Randel Morrison, Jeff Pew, Tom Ruhe, Kirsten Solder, Luke Sosnowski, Wei-Lit Teoh, Kevin Wang, and Fred Williams. It was through the use of Lean Development and discussions about implementation over the course of 16 years that formed his understanding of how to do the work differently. Lastly, Bob would like to thank his family, and specifically his wife Deb, for their dinnertime discussions about product development.

Alissa would like to thank her residency leadership, co-residents, mentors, and supervisors for providing a supportive, collegial, and educational learning environment and helping her develop into the physician she is today. She would additionally like to thank Nicole Helmke for her moral support during weekly walks as well as Rose Werth for providing constructive critiques. Finally, Alissa would like to thank Nate Wilson for his willingness to serve as an external motivator and task manager during this project.

Finally, we would like to thank the nearly 30 women who devoted their time through interviews which added personal perspectives to a very challenging topic.

REFERENCES

1. V. Valian, *Why so Slow? The Advancement of Women.* Cambridge, MA: MIT Press, 1998.
2. J. K. Fletcher, *Disappearing Acts: Gender, Power, and Relational Practice at Work.* Cambridge, MA: MIT Press, 1999.

Introduction

Centuries ago, the lives of sailors and the fate of their ships were dependent on their ability to navigate the oceans based on the stars and moon. However, stories of ships crashing into landmasses they sought to avoid or missing landmasses they hoped to find abound in seafaring history. Just like sailors were reliant on the stars, women engineers today rely on finding a supportive firm to navigate their careers successfully. They are looking for a firm with a work environment in which there are opportunities for growth and development, they are empowered to make engineering a lifelong career, and the diversity that they bring to the organization is valued.

While there have been significant efforts to improve the work environment and increase the number of women engineers, the percentage of women in engineering only increased from 12% in 1990 to 15% in 2019 [1]. Despite initiatives like addressing hiring practices and having trainings on implicit bias, the needle hasn't moved very far over the past three decades. These results indicate that something needs to change particularly as there are other STEM-related fields who have not encountered the same problem. In the field of medicine, women crossed the threshold and made up over 50% of students in medical schools in 2019 [2]. During that same year, women made up 36% of practicing physicians [3]. Why is there such a stark difference between these two fields of science? This book will attempt to address that question, using the field of medicine as a case study for the field of engineering to draw upon as it works to increase the retention of women engineers.

The genesis of this book came from the confluence of three distinct events. The first event occurred during the 2019 International Women's Week where the message was clearly communicated that individual male technology managers were responsible for creating a positive work environment for women by appropriately recognizing women's contributions and capabilities – a message that has been communicated for decades without the desired outcomes. At the same time, Alissa was in the second year of her residency training program (a physician's first job after medical school). Bob and Alissa noticed how dramatically different her learning environment was compared to that of most early-to-career engineers. One noticeable difference was that her access to web-based knowledge accounted for a significant portion of her knowledge-based learning. In contrast, an early-to-career engineer might have access to only a small portion of his/her learning needs and be entirely dependent on learning from others for the remainder of the information. The third event was Bob internalizing that if access to available knowledge was a key element in enabling the success of women in medicine, then Lean Development may be a path to address the knowledge gap within engineering given that a foundational element of Lean Development is the creation of reusable knowledge. With Bob's 15 years of utilizing the principles of Lean Development across a broad set of product and process delivery efforts and seeing positive outcomes, the idea that implementing

Lean Development processes could increase the success of women in engineering, in addition to the broader business benefits, was very compelling.

This book approaches the problem of women leaving engineering from an engineering problem-solving approach and outlines specific ways to change the way engineering is done while improving the overall competitiveness of an engineering firm. The book describes the challenges women face within engineering, draws from the learnings of training in medicine, and utilizes the six principles of Lean Development to change the engineering system. Integrated within the book are "gray box stories" to describe two different worlds that engineers could work in: a traditional world and the world of Lean Development. These stories become foundational elements for changing the work of engineering.

THE PRINCIPLES OF LEAN DEVELOPMENT – CHANGE THE SYSTEM OF ENGINEERING

For our use of Lean Development, we use six principles* that were first articulated 20 years ago [4].

1. The Creation of Reusable Knowledge
2. Cadence, Pull, and Flow
3. Visual Management
4. Entrepreneurial System Designer
5. Set-Based-Concurrent Engineering
6. "Build Teams of Responsible Experts"[1] [4, p. 1]

Additionally, we identify three key elements within a work environment that contribute to overall job satisfaction and work climate: how people treat each other, how the work is done, and what the work is. Using the six principles of Lean Development and the three key elements of a work environment, we create an approach to change the engineering system with the goal of creating a level playing field for women.

WHY THIS BOOK?

This book will take a new, engineering system-based approach to address the low retention rate of women in engineering by changing the learning process and how the work is done. This approach will result in changes in many areas including increased work satisfaction, improved work climate, personal growth, increased confidence, and decreased bias. Ultimately, we believe this new approach will lead to increased retention of women in engineering. While we anticipate that these changes will be particularly beneficial for women because of the unique challenges they face in engineering, it is a system-based change and therefore all members of the firm will experience the benefits. Further, given that a decade of research has shown that diverse

* Allen C. Ward identified five of these principles in his 2002 Lean Development Skills Book, Ward Synthesis, Inc. Visual Management has been added since that time.

work groups deliver better business results, increasing diversity in engineering will result in increased profit and revenue for the firm, fueling its growth [5].

Of note, a lack of women in engineering is not the only challenge to be addressed within the field of engineering as it has also yet to address the challenges faced by people marginalized on the basis of their race/ethnicity and sexuality. In particular, Black and Hispanic individuals remain underrepresented in engineering [1]. A Black woman's experience in engineering is therefore shaped by both race and gender resulting in an additional set of challenges. Each of these areas is worthy of focus and deserve time dedicated to them. While the scope of this book is limited to women in engineering, our hope is that the solutions discussed may also serve to address overall diversity in the field of engineering and improve outcomes for all groups who have been marginalized.

A BACK DROP WITHIN THE BOOK*

Europe's 18th century effort to "find the longitude" (a ship's east to west location while at sea) provides us with a back drop and a framework for the book. For centuries, the scientific community believed that the only solution to finding the longitude was from the stars and the moon. For decades, individuals looked to the stars to find the longitude while simultaneously creating hurdles for any other opposing solutions.

Eventually, the competing and effectively winning solution to the longitude challenge emerged from an entirely new direction in the form of the marine chronometer, a clock. As the book works through solving our problem, we too will find our clock. Surprisingly, just like a marine clock which you can hold in your hands and made its way to every ship, the clock we find can be held in your hands and can make its way to every engineer and leader.

REFERENCES

1. R. Fry, B. Kennedy, and C. Funk, "STEM Jobs See Uneven Progress in Increasing Gender, Racial and Ethnic Diversity," Pew Research Center, Apr. 2021. [Online]. Available: https://www.pewresearch.org/science/wp-content/uploads/sites/16/2021/03/PS_2021.04.01_diversity-in-STEM_REPORT.pdf.
2. P. Boyle, "More Women Than Men are Enrolled in Medical School," *AAMCNews*, Dec. 09, 2019. Accessed: Jan. 30, 2022. [Online]. Available: https://www.aamc.org/news-insights/more-women-men-are-enrolled-medical-school.
3. P. Boyle, "Nation's Physician Workforce Evolves: More Women, a Bit Older, and Toward Different Specialties," *AAMCNews*, Feb. 02, 2021. Accessed: Jan. 15, 2022. [Online]. Available: https://www.aamc.org/news-insights/nation-s-physician-workforce-evolves-more-women-bit-older-and-toward-different-specialties.
4. A. C. Ward, *The Lean Development Skills Book.* Ann Arbor, MI: Ward Synthesis, Inc., 2002.

* The historical context for this section comes from Dava Sobel's book *Longitude – The True Story of a Lone Genius Who Solved the Greatest Problem of His Time*, Penguin Books, 1995 [6]

5. V. Hunt, L. Yee, S. Prince, and S. Dixon-Fyle, "Delivering Through Diversity," McKinsey & Company, Report, Jan. 2018. Accessed: Jan. 15, 2022. [Online]. Available: https://www.mckinsey.com/business-functions/people-and-organizational-performance/our-insights/delivering-through-diversity.
6. D. Sobel, *Longitude: The True Story of a Lone Genius Who Solved the Greatest Scientific Problem of His Time*. New York, NY: Penguin Books, 1996.

Part I

The Problem

Women make up only 15% of engineers.* Two decades of effort has barely moved the needle.

* Pew Research Center. STEM Jobs See Uneven Progress in Increasing Gender, Racial and Ethnic Diversity. April 1, 2021. The report shows that women made up 12% of the engineering workforce in 1990 – resulting in only a 1% increase per decade.

DOI: 10.1201/9781003205814-1

1 The Female Engineer* Retention Challenge

Recent research shows that women make up only 15% of engineers in the workforce and 22% of engineering graduates with bachelor's degrees [1]. In addition, the Society of Women Engineers (SWE) Fast Facts report that "only 30% of women who earn bachelor's degrees in engineering are still working in engineering 20 years later" [2, p. 2]. Many studies have examined the overall career trajectories of engineers and assessed underlying causes that may account for differences between genders. Some studies found that balancing work and family life continues to be a significant factor, while others noted the contribution of negative work culture and the overall social structure of engineering [3–6]. A 2011 report of survey data from 3,700 women who had graduated with an engineering degree described issues such as lack of clear, tangible paths to advancement, unclear responsibilities and expectations, and the overall workplace climate [6].

FACTORS INHIBITING GIRLS AND WOMEN FROM CHOOSING ENGINEERING

The pipeline of women to engineering schools is no less challenged than retention. Bringing Science, Technology, Engineering, and Math (STEM) into the classroom is an ongoing and global effort. However, with so few women engineers as role models, what is the overall gain when a middle-aged white male (e.g., Bob) visits an eighth-grade science class or speaks to a group of third graders about "what does an engineer do?" Some girls may see past who the presenter is, but others may find the male presenter's sheer presence reinforcing the stereotype that engineering is for boys/men during the presentation and then hear that from classmates later on the playground.

There are dozens of studies completed each year to understand why women do not enter many STEM (Science, Technology, Engineering, and Math) fields at the same rate as men. Each study begins with a logically correct premise or hypothesis to test, focusing on understanding the main drivers for the typical stereotype or work climate issues and proposing changes to address them.† In SWE's State of Women in Engineering 2020 report, the authors write that "there is a growing sense in the

---

* For the purpose of this book, we use the phrase retaining women or female engineers to include women engineering managers. We also use the term women or female engineers to include all engineers who identify as female or women.

† For a full review of the most recent studies conducted, please see SWE's State of Women in Engineering 2020 report, which includes a 2019 literature review [7].

DOI: 10.1201/9781003205814-2

literature that women choose not to enter engineering and other STEM fields not because they have a strongly negative view of STEM, but because they find other fields more appealing" [7, p. 13]. As more occupations have become available to women over the past few decades, there are simply more options. Fields viewed as more welcoming of women have a more equal distribution of men and women and a better work climate. They are, therefore, naturally more appealing to women.

In addition to feeling welcome and supported in the workplace, studies have found the importance of an individual's attitudes in determining entrance into a STEM field, including attitudes about math skills and feelings of "fit" [8–11]. These attitudes can vary from having a falsely low belief in one's math ability to having high confidence in one's ability but finding more utility and value in other fields. Ultimately, the decision to enter (or not to enter) the engineering field is driven by a complex set of factors at both the individual and the broader occupational levels. Given the self-perpetuating cycle of engineering being less enticing due to unequal gender distribution, one key target to increasing the number of women entering the engineering field may be to address the issue of retention.

THE EFFORTS OF PROGRESSIVE FIRMS

Virtually every technology company has been working to attract and retain female engineers for years, if not decades. For those of us who have been in the industry during this process, we have watched and supported these efforts firsthand. Senior management has put time and energy into creating a balanced workforce. We understand the reasoning for it – there is benefit in having diverse perspectives when designing a solution to meet customer needs. As the McKinsey & Company states in their 2018 report *Delivering through Diversity,* "we found a positive correlation between gender diversity on executive teams and both our measures of financial performance. Top-quartile companies were 21% more likely than fourth quartile companies to outperform national industry peers on EBIT margin" [12, p. 10]. While they note that correlation does not indicate causation, there is a strong relationship between business outcomes and diversity [12]. A diverse executive staff can take the actions needed to retain female engineers resulting in increased diversity at the technical levels which then drives financial growth and creates a pipeline for ongoing diversity at the executive level. Simply put, having gender diversity makes good business sense.

Over the last two decades, initiatives to address male dominance in engineering have included recruiting methods, hiring practices, training in implicit bias, improving the work climate, encouraging women to seek promotional opportunities, and creating specific women leadership programs. Throughout this book, we will refer to firms that are actively trying to address the engineering gender imbalance issues as "progressive firms." These progressive firms are providing leadership by creating a positive work climate for women and, in general, are the firms most able to attract women to their ranks and retain them. The *Forbes* 2021 list of "America's Best Employers for Women" lists the top 300 companies and includes approximately 33 from the area of IT, engineering, or manufacturing [13]. These 33 companies range in size from a few thousand employees to over 100,000. The list contains companies that are household names as well as those that may only be known to individuals in that specific industry. These 33 companies and other progressive firms are working to understand the causes of the imbalance and put specific programs in place. While the engineering numbers for female engineers at many of these firms may be near the industry average, their management and leadership ranks may approach 50% women.

Progressive firm's efforts to improve work climate typically focus on work–life balance, fostering respect, increasing feelings of being valued, and accepting a diversity of personalities and work styles. In general, these areas address how people treat or speak to one another. However, implicit bias and the gender confidence gap can inhibit growth and improvement within each of these important areas. Implicit bias describes the unconscious perceptions or beliefs we may have about individuals which can then impact hiring and promotion [14]. The gender confidence gap, or the difference between men and women in perceived ability to complete a task or job, has been heavily researched and described [15, 16]. Implicit bias and gender confidence are engrained in our society and begin at an early age for most boys and girls. We will explore the systemic factors contributing to implicit bias and the gender confidence gap in Chapter 2.

Although a progressive firm can work to mitigate the effects of implicit bias within the work environment, they have little control outside the work environment. In addition to focusing on the work climate, the progressive firm may be working toward establishing a structure of work that can benefit women, including grouping women together versus spreading them thinly across teams. However, this intervention may have unintended consequences like decreasing an overall sense of belonging in the field if women are isolated to a specific team, and creating an undesirable differentiation across a firm's engineering staff [7]. Ultimately, these methods aimed at targeting the characteristics and attitudes of women on an individual level or addressing the overall work climate have not been effective in significantly increasing the number of women in engineering and reducing the difference in retention rate.

[Of note, there are even more challenges for women marginalized on the basis of their race/ethnicity or sexuality in addition to their gender. These areas are worthy of specific focus and deserve further time dedicated to them. While the scope of this book is limited to women in engineering, as we noted in the Introduction section, our hope is that the solutions discussed may also serve to address overall diversity in the field of engineering.]

FINDING A NEW APPROACH

Rather than organization-wide interventions or targeting specific areas, both of which have not been successful thus far, we propose that it is time to change the system of engineering* itself. We believe that the way traditional engineering is done

* The term "system of engineering" includes every part that is required to deliver a solution to a customer. The system encompasses understanding the need, the necessary skill sets, the clarity of objectives, the viability of a solution, how the work is done, and the need for diverse thought processes.

impacts not only the retention of women but also the attraction of women to the profession. In the remainder of this book, we will use a system-based approach to break the problem down into its major components, work toward finding the root causes of unequal retention rates between men and women, and propose solutions focused on changing the system of engineering. Changing the work of engineering is not purely for the benefit of women but also for the benefit of a firm's competitiveness – namely their profit and revenue. In searching for ways to change the engineering system to facilitate increased gender balance, it is helpful to turn to STEM fields in which there has been success, particularly the field of medicine and view them as a case study.

APPROXIMATELY 50% OF THE MEDICAL SCHOOL GRADUATES ARE WOMEN; WHY?

While in the field of engineering there has been little increase in the percentage of women in the field over the past 20 years, in 2019 the male-dominated profession of physicians achieved a 50/50 male-to-female split of students enrolled in medical schools [17]. Medicine has seen a steady increase in the percentage of women in medical schools since the 1950s. This difference brings up the natural question of how has medicine been so successful in achieving these results? Indeed, a major contributor to achieving this 50/50 split is the overall high interest in entering medicine, as evidenced by the competitiveness of being admitted to medical school. In the 2018–2019 school year, only about 40% of applicants were accepted into medical schools [18]. With approximately 52% of applicants that year being women [18], medical schools had an excellent pipeline to enable gender diversity. The competitiveness of applicants allows medical schools to draw from top candidates and achieve a diverse class. Upon entry into medical school, those students are part of a structured teaching and learning process which continues after medical school graduation through entry into residency* (a physician's first paid position).

This then leads to the next question: What are the underlying factors that lead to an increased interest in pursuing medicine rather than engineering? There are likely many factors, including the importance of the humanities in medicine and the interpersonal aspect of the doctor–patient relationship. This may result in a broader array of individuals interested in medicine than individuals interested in engineering. Further, there is likely a positive influence from the number of female physician role models available as well as the number of female physicians portrayed in films and television. In contrast, there are fewer female engineers portrayed in media compared to male engineers [19].

---

* Residency is post-graduate training for qualified physicians (e.g., holding a degree of MD or DO), who practice medicine under the supervision of a senior physician. A residency program length varies by specialty but ranges from three to seven years. Residency may be followed by a one- to three-year fellowship, which provides additional specialty training. We will refer to residency and fellowship as graduate medical education.

Unlike engineering, medicine has a formal training process after leaving medical school – residency and fellowship. There are multiple stages along this path at which we can measure attrition. Per AAMC* data, the overall attrition rate in medical school is low (3% over the past 20 years), although that information is not broken down by gender [20]. While the attrition rate in residency is overall low, reported by the *ACGME*† *Data Resource Book* for 2019–2020 to be less than 4%, there is significant variability in the attrition rate between specialties [21]. While one 2008 study found that 6% of residents and fellows who were recent graduates of a single medical institution did not complete their training program and that there was not a significant difference between men and women [22], other studies of specific specialties, like surgical residents, have found significant differences between attrition of men and women [23]. Post-residency and post-fellowship, there are significant gender differences between part-time work and full-time work, with one study finding a significantly higher percentage of women working part time (22.6%) compared to men (3.6%) [24]. While the study did not report the gender breakdown of physicians who were no longer working in medicine, the number of physicians in that study leaving the field of medicine entirely was overall very low (2% or seven individuals) [24]. Qualitative data completed during the study found that the primary factor contributing to an individual's decision to transfer to part-time employment was work–family conflict. Notably, if a similar study was completed in 2022 in the context of the global COVID-19 pandemic, these numbers may look very different.

While many factors may arise after completing graduate medical education that contribute to early career female physicians transitioning to part-time employment at a much higher rate than men, including lack of family leave support, during residency itself there is a formal training process that provides structured learning to residents regardless of gender. A physician residency program is typically completed immediately after graduating from medical school. The training is directly focused on a resident's desired specialty. Salaries are standardized at each institution based on the year of training. Further, there is a national accrediting body (ACGME), which is responsible for overseeing the accreditation of each program, ensuring specific standards are met and carrying out an annual survey of residents. While issues remain within the structure of the system, particularly surrounding family leave policies, many areas of potential bias that may occur in an unstructured learning and work environment are reduced.

There are notable differences between the fields of medicine and engineering, including the necessity of completing a residency in order to practice in a particular specialty, the types of individuals who may be drawn to medicine, the view of medicine as a vocation rather than a career, and the fundamental importance of the doctor–patient relationship. However, there are also notable similarities, including a foundation in scientific knowledge, the importance of critical thinking, and the need to adapt and innovate to address problems. Perhaps most importantly, though, prior approaches to increase the number of women in engineering have not been effective

* AAMC – American Association of Medical Colleges.

† ACGME – Accreditation Council for Graduate Medical Education.

and we propose that it is now time to look at other successful fields for guidance. Given the relatively low attrition rate in residency in comparison to the attrition rate in engineering, the process of teaching and training in medicine has valuable lessons for training new engineers.

CONCLUSION

We have laid out the problem facing women in engineering and the field of engineering as a whole: despite decades of effort, women make up only 15% of engineers and 22% of those graduating with engineering degrees. Given the lack of success that firms have had in trying to address this problem through improving the work climate and increasing opportunity for advancement, it is time to look for a new system-based approach and gain ideas from other fields, specifically the field of medicine.

The rest of this book will break down the gender imbalance in engineering into three basic components: how people interact, how the work is done, and what the work is. We will identify specific changes that can be made within the system of engineering to level the playing field for female engineers. To do this, we will use the guiding question, "What can the field of engineering learn from the field of medicine as it relates to retaining women, and what should we change within engineering?"

REFERENCES

1. R. Fry, B. Kennedy, and C. Funk, "STEM Jobs See Uneven Progress in Increasing Gender, Racial and Ethnic Diversity," Pew Research Center, Apr. 2021. [Online]. Available: https://www.pewresearch.org/science/wp-content/uploads/sites/16/2021/03/PS_2021.04.01_diversity-in-STEM_REPORT.pdf.
2. Society of Women Engineers, "SWE Research Fast Facts," Society of Women Engineers (SWE), Sep. 2021. [Online]. Available: https://swe.org/wp-content/uploads/2021/10/SWE-Fast-Facts_Oct-2021.pdf.
3. E. A. Cech, and M. Blair-Loy, "The Changing Career Trajectories of New Parents in STEM," *Proceedings of the National Academy of Sciences of the United States of America*, vol. 116, no. 10, pp. 4182–4187, Mar. 2019, doi: 10.1073/pnas.1810862116.
4. L. Ettinger, N. Conroy, and W. Barr II, "What Late-Career and Retired Women Engineers Tell Us: Gender Challenges in Historical Context," *Engineering Studies*, vol. 11, no. 3, pp. 217–242, Sep. 2019, doi: 10.1080/19378629.2019.1663201.
5. S. E. Khilji, and K. H. Pumroy, "We are Strong and we are Resilient: Career Experiences of Women Engineers," *Gender Work Organ*, vol. 26, pp. 1032–1052, 2019, doi: 10.1111/gwao.12322.
6. N. A. Fouad, and R. Singh, "Stemming the Tide: Why Women Leave Engineering," Center for the Study of the Workplace at University of Wisconsin - Milwaukee, 2011. [Online]. Available: https://www.energy.gov/sites/prod/files/NSF_Stemming%20the%20Tide%20Why%20Women%20Leave%20Engineering.pdf.
7. P. Meiksins, P. Layne, K. Beddoes, and J. Deters, "Women in Engineering: A Review of the 2019 Literature," *SWE Magazine*, vol. 66, no. 2, pp. 4–41, Apr. 2020.
8. E. Seo, Y. Shen, and E. C. Alfaro, "Adolescents' Beliefs about Math Ability and Their Relations to STEM Career Attainment: Joint Consideration of Race/Ethnicity and Gender," *Journal of Youth and Adolescence*, vol. 48, no. 2, pp. 306–325, Feb. 2019, doi: 10.1007/s10964-018-0911-9.

9. A. Heyder, A. F. Weidinger, and R. Steinmayr, "Only a Burden for Females in Math? Gender and Domain Differences in the Relation Between Adolescents' Fixed Mindsets and Motivation," *Journal of Youth and Adolescence*, vol. 50, no. 1, pp. 177–188, Jan. 2021, doi: 10.1007/s10964-020-01345-4.
10. M.-T. Wang, and J. Degol, "Motivational Pathways to STEM Career Choices: Using Expectancy–Value Perspective to Understand Individual and Gender Differences in STEM Fields," *Developmental Review*, vol. 33, no. 4, pp. 304–340, Dec. 2013, doi: 10.1016/j.dr.2013.08.001.
11. D. Verdín, A. Godwin, A. Kirn, L. Benson, and G. Potvin, "Engineering Women's Attitudes and Goals in Choosing Disciplines with Above and Below Average Female Representation," *Social Sciences*, vol. 7, no. 3, pp. 44, Mar. 2018, doi: 10.3390/socsci7030044.
12. V. Hunt, L. Yee, S. Prince, and S. Dixon-Fyle, "Delivering Through Diversity," McKinsey & Company, Report, Jan. 2018. Accessed: Jan. 15, 2022. [Online]. Available: https://www.mckinsey.com/business-functions/people-and-organizational-performance/our-insights/delivering-through-diversity.
13. V. Valet, "Best Employers for Women," *Forbes*, Jul. 27, 2021. Accessed: Feb. 18, 2022. [Online]. Available: https://www.forbes.com/best-employers-women/#1d90c8f77de9.
14. C. Isaac, B. Lee, and M. Carnes, "Interventions That Affect Gender Bias in Hiring: A Systematic Review," *Academic Medicine*, vol. 84, no. 10, pp. 1440–1446, Oct. 2009, doi: 10.1097/ACM.0b013e3181b6ba00.
15. C. Exley, and J. Kessler, "The Gender Gap in Self-Promotion," National Bureau of Economic Research Working Paper Series, vol. No. 26345, Oct. 2019, doi: 10.3386/w26345.
16. J. Ellis, B. K. Fosdick, and C. Rasmussen, "Women 1.5 Times More Likely to Leave STEM Pipeline after Calculus Compared to Men: Lack of Mathematical Confidence a Potential Culprit," *PLoS ONE*, vol. 11, no. 7, pp. e0157447, Jul. 2016, doi: 10.1371/journal.pone.0157447.
17. S. Heiser, "The Majority of U.S. Medical Students are Women, New Data Show," *AAMCNews*, Dec. 09, 2019. Accessed: Jul. 18, 2021. [Online]. Available: https://www.aamc.org/news-insights/press-releases/majority-us-medical-students-are-women-new-data-show.
18. Association of American Medical Colleges (AAMC), "2019 Fall Applicant, Matriculant, and Enrollment Data Tables," Association of American Medical Colleges (AAMC), Dec. 2019. Accessed: Jul. 05, 2021. [Online]. Available: https://www.aamc.org/media/38821/download.
19. The Lyda Hill Foundation & The Geena Davis Institute on Gender in Media, "Portray Her: Representations of Women STEM Characters in Media," The Lyda Hill Foundation & The Geena Davis Institute on Gender in Media, Online, c 2022. Accessed: Jan. 15, 2021. [Online]. Available: https://seejane.org/wp-content/uploads/portray-her-full-report.pdf.
20. Association of American Medical Colleges (AAMC), "Graduation Rates and Attrition Rates of U.S. Medical Students," Association of American Medical Colleges (AAMC), AAMC Data Snapshot, Oct. 2018. Accessed: Jan. 31, 2022. [Online]. Available: https://www.aamc.org/system/files/reports/1/graduationratesandattritionratesofu.s.medical-students.pdf.
21. Accreditation Council for Graduate Medical Education, "Data Resource Book: Academic Year 2020–2021," Accreditation Council for Graduate Medical Education, Chicago, IL, Data Resource Book, 2021. Accessed: Dec. 20, 2021. [Online]. Available: https://www.acgme.org/globalassets/pfassets/publicationsbooks/2020-2021_acgme_databook_document.pdf.

22. D. A. Andriole, D. B. Jeffe, H. L. Hageman, M. E. Klingensmith, R. P. McAlister, and A. J. Whelan, "Attrition During Graduate Medical Education: Medical School Perspective," *Archives of Surgery*, vol. 143, no. 12, pp. 1172–1177, Dec. 2008, doi: 10.1001/archsurg.143.12.1172.
23. Z. Khoushhal *et al.*, "Prevalence and Causes of Attrition Among Surgical Residents: A Systematic Review and Meta-Analysis," *JAMA Surgery*, vol. 152, no. 3, pp. 265–272, Mar. 2017, doi: 10.1001/jamasurg.2016.4086.
24. E. Frank, Z. Zhao, S. Sen, and C. Guille, "Gender Disparities in Work and Parental Status Among Early Career Physicians," *JAMA Network Open*, vol. 2, no. 8, pp. e198340, Aug. 2019, doi: 10.1001/jamanetworkopen.2019.8340.

2 Two Sides of the Same Coin
Gender Bias and Gender Confidence

Over the past several decades, there have been many studies done, articles published, and books written about bias and confidence in the workforce and the ways in which they affect the success of women in male-dominated fields. This chapter aims to provide a brief overview of these topics and how they relate to the retention of female engineers. Given the breadth of the subject and volume of work written, we cannot fully explore all relevant studies, articles, and books in this section; however, we refer you to our citations for more in-depth exploration.

GENDER SCHEMA AND STEREOTYPES

We all carry perceptions and assumptions about the world around us based on our own life experiences, interactions with others, and implicit and explicit messages from the broader culture and society we live in [1]. These assumptions can be beneficial in our day-to-day lives – they allow us to generalize specific experiences to other situations and make quick decisions. For a basic example, a child who burns themselves after touching a red-hot stove may generalize from this experience and hypothesize (correctly) that red-hot items may cause pain. The next time they encounter a red-hot item, they may automatically hesitate to touch it. In clinical medicine, we use associations and diagnostic schema to categorize clinical information and make clinical decisions. They help us decide what diseases are most likely present and what further testing we need.

However, associations about people, based on characteristics like race, age, and gender, are much more complicated and can result in biased decision-making even in those who express egalitarian beliefs [1, 2]. The psychologist Virginia Valian uses the term gender schema to describe these associations based on gender. Specifically, in *Why So Slow? The Advancement of Women*, she defines gender schema as "a set of implicit, or nonconscious, hypotheses about sex differences" [1, p. 2] which "affect our expectations of men and women, our evaluations of their work, and their performance as professionals" [1, p. 2]. She uses the term schema, rather than stereotypes, as a broader way of describing how we categorize the world around us. While the term gender schema has also been used in other ways, specifically to describe how individuals impose expectations related to gender on themselves, we are using the

DOI: 10.1201/9781003205814-3

term "gender schema" to describe the general hypotheses we have about men and women [1].

Gender schema includes both assumptions that can be objectively measured, like men being taller, and subjective characteristics, like men being competent and assertive while women are warm and nurturing [1, 2]. In a 2018 review, Ellemers summarizes the evidence regarding our beliefs about men and women and describes how these beliefs shape the behavior of men and women over time and have long-term career implications [2]. Studies of individuals rating identical job applications have shown a difference in perceived competency if the applicant was a male or female [3] and a significant difference in job rewards, like salary and promotion, that was not explained by performance evaluation differences [4]. Further, men and women have overall been shown to be much more similar than they are different, with more significant variability within each group than between the two groups [1, 2, 5].

Of note, throughout this book, we use the term sex to describe the label assigned based on biological characteristics and the term gender to describe the associated social and cultural traits and differences [6]. As described by Bussey, "[g]ender identity is informed by knowledge of one's biological sex and of the beliefs associated with gender, how one is perceived and treated by others depending on one's gender, and an understanding of the collective basis of gender" [6, p. 608]. Gender identity is therefore not purely linked to sex and can also differ from gender expression or how an individual outwardly presents their gender. While the ways in which specific forms of gender identity and gender expression intersect with experiences in engineering are worthy of additional consideration and exploration, they are outside of the limited scope of this book. For our purposes, we will use the term gender broadly, recognizing that the expression of gender (i.e., masculinity and femininity) exists on a spectrum rather than being binary and determined solely by biology [1, 6].

As the way we talk about gender expression, gender identity, sex, and gender roles has changed, many parents are focusing on raising their children in a gender-neutral way [7]. While this has shifted the explicit messages children receive from their parents, recent studies have shown that gender schemas remain set and are resistant to change. A 2021 study examined individuals' responses to a riddle that required them to recognize the possibility that a woman could be a surgeon [8]. Only 30% were able to think outside of the traditional gender role of a male surgeon and consider that a woman surgeon could be the correct answer [8]. This occurred even though the number of women physicians and women surgeons has increased over time.

While schema can be helpful, there are also significant risks of bias. Having a rigid and fixed schema can result in confirmation bias, a tendency to interpret experiences and interactions in a way that confirms what we already believe to be true [9]. For example, if a male student performs well on a math test and we have a perception that boys are better at math, we may accept this as an example of his natural ability and a sign that he is "smart" [1, 2]. In contrast, if a female student does well on a math test, and we have a perception that girls are not good at math, we may come up with explanations such as she put forth a lot of effort or this was an easy exam rather than changing our hypothesis and viewing her as having math ability [2]. This demonstrates our tendency to maintain our initial beliefs rather than integrate new and contradictory experiences [10], also known as belief perseveration bias. Gender schema can therefore lead to specific gender biases which affect how we interact with and respond to others [1], leading us to treat men and women differently, perpetuating and increasing differences between men and women [2].

GENDER BIAS – THE EXTERNAL INFLUENCER

As discussed, unconscious and conscious bias exists in us all and gender bias specifically arises from gender schema, or the hypotheses we have about men and women. Importantly, in thinking about gender bias in the workplace, men are overrated and viewed as more competent than they are, while women are underrated and viewed as less competent than they are [1, 3, 11]. Further, studies have shown that people implicitly associate men with careers and women with family regardless of the participant's gender [2, 12]. Similar biases exist regarding scientific and mathematical abilities, with one study demonstrating that men were two times more likely than women to be hired to perform an arithmetic task [13]. Like with other studies, this difference existed regardless of the gender of the employer [13]. It is not surprising then that our biases affect hiring, promotion, and assignment of responsibilities at work [4, 11, 14, 15]. Both the advantages men experience and the disadvantages women experience accumulate over time, widening the gap [1, 2, 16].

Significant work has been done to address unconscious bias – both to better understand it and to recommend ways for employers to minimize the effects of gender-related bias.

In *The End of Bias: A Beginning: The Science and Practice of Overcoming Unconscious Bias*, Nordell describes five specific ways in which bias manifests in the workplace [16, p. 80–82]:

- Women's performance is not valued as highly as men's performance.
- Women's errors result in a higher penalty than men's errors.
- Fewer opportunities are provided to women than men, termed opportunity bias.
- Loss of credit for contributions to work.
- Negative consequences for women with personality traits that are not in line with societal expectations for women.

To better understand bias in engineering, in 2016, the Center for WorkLife Law and Society of Women Engineers (SWE) published results from a survey of over 3,000 respondents, identifying significant gender gaps in three patterns of bias [17]:

- "Prove it again": 61% of women report that to get the same level of respect and recognition as their colleagues, they need to repeatedly prove themselves. This was in comparison to just 35% of white men reporting the same.
- "Tightrope Bias": There is a more limited range of acceptable behavior for women in comparison to men. For example, the author writes that "[w]omen engineers were less likely than white men to say they could behave

assertively (51% vs. 67%) or show anger without pushback (49% vs. 59%)" [17, p. 3].

- "Maternal Wall": While only 55% of women said that having children did not change how their commitment to work was viewed, almost 80% of men said the same.

These biases exist not only on an individual level but also on a broader system level through standardized processes like performance evaluations [18, 19]. Addressing unconscious bias is, therefore, multifactorial and dependent not only on individuals identifying their own biases, and being aware of them when making decisions, but also upon companies making changes to standardized processes in order to address implicit biases inherent within the system [15]. The gender bias that women experience within engineering influences their desire to remain within the field.

GENDER CONFIDENCE – THE INTERNAL INFLUENCER

Merriam-Webster has defined confidence as "a feeling or consciousness of one's powers or of reliance on one's circumstances" [20]. Differences in confidence, or perceived ability to complete a task or job, have been demonstrated between men and women, and women routinely underestimate their abilities compared to men [21, 22]. These differences in perceived abilities begin in adolescence and shape career trajectories from a young age. In a study by Christine L. Exley and Judd B. Kessler of over 4,000 participants, they found that "when asked to indicate agreement on a scale from 0 to 100 with a statement that reads 'I performed well on the test,' women provide answers that are 13 points lower than equally performing men"* (or a 24% gender gap) [23, p. 2]. The term gender confidence gap describes the difference in confidence between the two genders.

The Confidence Code (2014), written by Katty Kay and Claire Shipman, provides a detailed look at confidence, differences in confidence between genders, and the factors that contribute to the development of confidence – describing confidence as "the stuff that turns thoughts into action" [24, p. 50]. Kay and Shipman note how confidence develops over time and how views of failure and willingness to take risks differ between the genders. They note that a tendency toward perfectionism and overpersonalization of failure can inhibit women from taking action, speaking up, and putting themselves forward [24]. Women have been noted to be less likely than men to apply for a job when they don't fully meet the requirements [24, 25].

Research on gender confidence by Zenger Folkman has shown that women's confidence levels do ultimately catch up to men's. However, women in their mid-twenties have only 65% of the confidence men of a similar age have and do not achieve parity until their mid-forties [26]. During that span of 20 years, women make decisions about careers and career trajectories, are promoted (or not), and women begin leaving engineering at a higher rate than men. By the time women achieve parity in confidence to men, the damage has already been done.

---

* © Christine L Exley and Judd B. Kessler.

In *The Confidence* Code, Kay and Shipman provide concrete ways to build or maintain confidence. Notably, these are through avenues that allow for risk-taking and building resilience in supportive environments. In addition, the authors encourage taking action, leaving your comfort zone, building a positive mindset, and embracing failures [24].

BUILDING CONFIDENCE

In reflecting on her trajectory thus far, the idea of perfectionism and fear of failure is something Alissa relates to well. She recalls the times she avoided offering an answer or an opinion in a class or group as she was not 100% certain of the answer and was afraid of what might have occurred if she were incorrect. She remembers the projects and ideas she didn't pursue because she felt that others were more qualified. In reality, though, providing an incorrect answer would have resulted in learning something new, rather than being a moral reflection of her character, and would have allowed for ongoing strengthening of a muscle we can define as "confidence." Pursuing projects she was passionate about, while outside of her comfort zone, could have led to new opportunities or may not have turned out well. Regardless of the outcome, the act of pursuing them would have been more in line with Alissa's values of ongoing self-growth rather than following a path of inaction driven by a desire to avoid feelings of failure. Being surrounded by mentors who share their challenges, failures, and feelings of being imposters and having supportive colleagues have provided Alissa a safe environment to recognize her strengths, take risks, and step outside of her comfort zone – including co-authoring this book.

TWO SIDES OF THE SAME COIN – BIAS AND CONFIDENCE

Bias and confidence are interrelated. In a male-dominated group setting, a woman's hesitancy to speak may be due to a lack of confidence. Alternatively, it may be driven by an awareness of implicit biases and a desire to avoid being considered "too assertive" or avoid proving someone's hypothesis "right" that women are not as competent as men by having the wrong answer [1, 2]. In many cases, it is hard, and potentially impossible, to determine the underlying cause as both a lack of confidence and a reasonable, rational response to bias can result in a similar outcome. It is more helpful to reflect on how they relate to each other and the individual and address them simultaneously.

We consider bias an external influencer because it is driven by the interactions with the people around an individual. Similarly, we consider confidence an internal influencer because, as noted earlier, "Confidence is stuff that turns thoughts into action" [24, p. 50]. However, experiencing bias is frequently internalized and can affect confidence. With this interconnectedness, Figure 2.1 shows a simple model to demonstrate their interactions.

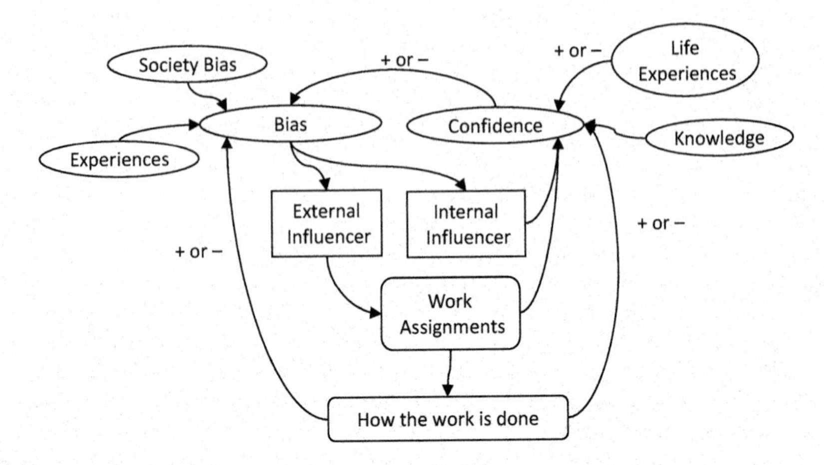

FIGURE 2.1 Implicit bias – confidence model (+ or –, can either be a positive or negative influence).

In the previous section, we saw that women in their mid-twenties, on average, start out at a confidence deficit compared to men of a similar age. This confidence deficit, in combination with gender bias, may result in a significant challenge to overcome. As discussed earlier, an individual's confidence is influenced by life experience (i.e., making mistakes, recovering from failures, developing resilience) and knowledge. Researchers have promoted the need for girls to "tinker" with stuff. This not only gives them knowledge about how things work and how to take something apart and put it back together again but is also an opportunity to try things out, experience obstacles, potentially fail, and then practice overcoming them [24].

This gives us a place to start with increasing confidence, focusing on areas that an individual can control, at least on the surface. These core components are summarized as follows.

Confidence – From Knowledge

Having a strong foundation of knowledge can enhance confidence; however, having the requisite knowledge alone is not enough to ensure confidence.

Confidence – From Experiences

Experiences provide the necessary training for taking chances. Experiences could be divided into two general categories:

1. Physical – e.g., activities like sports that one engages in or a healthy work team that can foster opportunities to make mistakes and learn from them
2. Emotional – relationships with others which can be positive or negative, permissive or authoritative

As we think about what individuals can do to increase their confidence, knowledge and experience are key areas that can be encouraged and developed.

In Figure 2.1, we show that an individual's bias comes from societal norms and personal experiences. An individual's interactions with others can serve to reinforce, confirm, reduce, or challenge bias. In our diagram, if an individual has a preconceived perception that another employee lacks competence, like what happens with men and women in the workplace, this can result in further bias. If that individual perceives that the employee appears to be confident during an interaction, this may challenge their hypothesis and perhaps reduce their bias. Alternatively, if they perceive that the employee lacks confidence, this can reinforce their prior belief. Notably, it is much harder to challenge a schema and overcome belief perseveration bias than to have an interaction serve as confirmation of our bias.

Bias can have two different effects: (1) external effects like marginal work assignments and lack of promotions; and, (2) internal effects like feelings of failure in response to comments about one's capability or ability to succeed. These internalized feelings can reinforce any negative self-perceptions, thus enhancing any underlying fear of failure or fear of not being perfect. Similarly, receiving a marginal work assignment can be perceived as confirmation by an individual already lacking in confidence that they are indeed correct to lack confidence and are indeed not competent.

The external effects of bias, like work assignments, can be obvious. An easy example would be a manager taking a week of vacation and asking one of his male engineers to lead a critical technical design review meeting in his absence while at the same time asking his only female engineer on the team, who is equally capable, to lead the team's staff meeting during that week. (No comment on whether she is also responsible for bringing snacks.)

HOW THE WORK IS DONE

Focusing on the individual, through efforts to reduce bias or increase confidence, has not resulted in the desired outcomes. Although we can, and should, have organizations work to reduce implicit bias, individual decisions and responses are difficult to fully influence and take time to change. Further, while books have been written about the concrete steps that individual women can take to increase confidence and better navigate the workplace [24, 27], given the systematic and pervasive nature of gender schema, it is also not reasonable nor sufficient to focus solely on women making individual changes. To put it plainly, just expecting someone to continually focus on making lemonade out of lemons is neither fair nor achievable.

We must therefore look for a broader system-based approach to simultaneously increase confidence and reduce bias. One approach is to look at the way the work is done, which can impact both confidence and bias (Figure 2.1). For example, if an individual is assigned a subpar work assignment but they are able to complete it effectively and competently, their confidence will likely increase and the bias of others may decrease. Importantly, how the work is done is within the control of an individual and can also be addressed through a system-based approach. We could imagine finding a system-based approach that increases access to knowledge as one of its methods to change the way the work is done and therefore resulting in increased confidence. Further, we could envision designing this system to also mitigate the effects of bias. This book will introduce a method of doing the work, specifically implementing Lean Development processes, which can result in increased knowledge, and therefore increased confidence, while mitigating the effects of bias.

CONCLUSION

Dozens of authors have published works about gender bias in male-dominated workplaces. In her 2021 book, *The End of Bias*, Jessica Nordell writes about a 2011 supreme court ruling concerning a class-action lawsuit that represented 1.6 million female Walmart employees. Per Nordell, the court concluded, "that daily biases do not add up to a large harm" [16, p. 72]. Nordell, however, notes that bias is experienced over weeks, months, and years and may affect a person's subsequent decisions and behaviors [16].

This chapter outlined the ways in which confidence and bias, resulting from our underlying gender schema, influence women's experiences at work and can have long-term career implications. Further, given the pervasive nature of gender schema and bias, as well as the ways that bias and confidence influence each other, we need a system-based approach rather than focusing on the individual. One way to change the overall system is to focus on how the work is approached. This can build a person's career-related confidence through knowledge and experiences, and mitigate the negative effects of bias.

In responding to the supreme court ruling noted earlier, Nordell proposes that "if we begin to perceive the workplace as a complex system, new insights become possible. To assess any true impact of bias, we must look not at the single moments but at the result of many, many, interactions" [16, p. 75]. Our hope is that this book is driven by new insights.

REFERENCES

1. V. Valian, *Why so Slow? The Advancement of Women.* Cambridge, MA: MIT Press, 1998.
2. N. Ellemers, "Gender Stereotypes," *Annual Review of Psychology*, vol. 69, pp. 275–298, Jan. 2018, doi: 10.1146/annurev-psych-122216-011719.
3. C. A. Moss-Racusin, J. F. Dovidio, V. L. Brescoll, M. J. Graham, and J. Handelsman, "Science Faculty's Subtle Gender Biases Favor Male Students," *Proceedings of the National Academy of Sciences of the United States of America*, vol. 109, no. 41, pp. 16474–16479, Oct. 2012, doi: 10.1073/pnas.1211286109.
4. A. Joshi, J. Son, and H. Roh, "When Can Women Close the Gap? A Meta-Analytic Test of Sex Differences in Performance and Rewards," *Academy of Management Journal*, vol. 58, no. 5, pp. 1516–1545, Oct. 2015, doi: 10.5465/amj.2013.0721.
5. J. S. Hyde, "Gender Similarities and Differences," *Annual Review of Psychology*, vol. 65, no. 1, pp. 373–398, Jan. 2014, doi: 10.1146/annurev-psych-010213-115057.

6. K. Bussey, "Gender Identity Development," in *Handbook of Identity Theory and Research*, vol. 1 & 2, S. Schwartz, K. Luyckx, and V. Vignoles, Eds. Berlin, Germany: Springer, Springer Nature, 2011, pp. 603–628.
7. M. Tortorello, "How to Raise a Child Without Imposing Gender," *The New York Times*, Mar. 07, 2019. https://www.nytimes.com/2019/03/07/style/gender-neutral-design-child.html (accessed Jan. 15, 2022).
8. D. Belle, A. Tartarilla, M. Wapman, M. Schlieber, and A. Mercurio, "'I Can't Operate, that Boy is my Son!': Gender Schemas and a Classic Riddle," *Sex Roles*, vol. 85, pp. 161–171, Jan. 2021.
9. R. S. Nickerson, "Confirmation Bias: A Ubiquitous Phenomenon in Many Guises," *Review of General Psychology*, vol. 2, no. 2, pp. 175–220, Jun. 1998, doi: 10.1037/1089-2680.2.2.175.
10. D. H. J. Wigboldus, A. Dijksterhuis, and A. van Knippenberg, "When Stereotypes Get in the Way: Stereotypes Obstruct Stereotype-Inconsistent Trait Inferences," *Journal of Personality and Social Psychology*, vol. 84, no. 3, pp. 470–484, 2003, doi: 10.1037/0022-3514.84.3.470.
11. J. F. Madden, "Performance-Support Bias and the Gender Pay Gap among Stockbrokers," *Gender & Society*, vol. 26, no. 3, pp. 488–518, Jun. 2012, doi: 10.1177/0891243212438546.
12. A. Salles *et al.*, "Estimating Implicit and Explicit Gender Bias Among Health Care Professionals and Surgeons," *JAMA Network Open*, vol. 2, no. 7, pp. e196545, Jul. 2019, doi: 10.1001/jamanetworkopen.2019.6545.
13. E. Reuben, P. Sapienza, and L. Zingales, "How Stereotypes Impair Women's Careers in Science," *Proceedings of the National Academy of Sciences of the United States of America*, vol. 111, no. 12, pp. 4403–4408, Mar. 2014, doi: 10.1073/pnas.1314788111.
14. C. Isaac, B. Lee, and M. Carnes, "Interventions That Affect Gender Bias in Hiring: A Systematic Review," *Academic Medicine*, vol. 84, no. 10, pp. 1440–1446, Oct. 2009, doi: 10.1097/ACM.0b013e3181b6ba00.
15. "Breaking Barriers: Unconscious Gender Bias in the Workplace," ACT / EMP The Bureau for Employers' Activities, Research Note, Aug. 2017. Accessed: Jan. 15, 2022. [Online]. Available: https://www.ilo.org/wcmsp5/groups/public/---ed_dialogue/---act_emp/documents/publication/wcms_601276.pdf.
16. J. Nordell, The End of Bias: A Beginning: The Science and Practice of Overcoming Unconscious Bias, First edition. New York: Metropolitan Books, Henry Holt and Company, 2021.
17. J. C. Williams, S. Li, R. Rincon, and P. Finn, "Climate Control: Gender and Racial Bias in Engineering?," Center for WorkLife Law & Society of Women Engineers, Executive Summary, 2016. Accessed: Dec. 03, 2021. [Online]. Available: https://worklifelaw.org/publications/Climate-Control-Gender-And-Racial-Bias-In-Engineering.pdf.
18. L. N. Mackenzie, J. Wehner, and S. J. Correll, "Why Most Performance Evaluations are Biased, and How to Fix Them," *Harvard Business Review*, Jan. 11, 2019. Accessed: Jan. 15, 2021. [Online]. Available: https://hbr.org/2019/01/why-most-performance-evaluations-are-biased-and-how-to-fix-them.
19. S. J. Correll, K. R. Weisshaar, A. T. Wynn, and J. D. Wehner, "Inside the Black Box of Organizational Life: The Gendered Language of Performance Assessment," *American Sociological Review*, vol. 85, no. 6, pp. 1022–1050, Dec. 2020, doi: 10.1177/0003122420962080.
20. "Confidence," *Merriam-Webster.com Dictionary*. Merriam-Webster. Accessed: Dec. 03, 2021. [Online]. Available: https://www.merriam-webster.com/dictionary/confidence.

21. E. Seo, Y. Shen, and E. C. Alfaro, "Adolescents' Beliefs about Math Ability and Their Relations to STEM Career Attainment: Joint Consideration of Race/ethnicity and Gender," *Journal of Youth and Adolescence*, vol. 48, no. 2, pp. 306–325, Feb. 2019, doi: 10.1007/s10964-018-0911-9.
22. J. Ellis, B. K. Fosdick, and C. Rasmussen, "Women 1.5 Times More Likely to Leave STEM Pipeline after Calculus Compared to Men: Lack of Mathematical Confidence a Potential Culprit," *PLOS ONE*, vol. 11, no. 7, pp. e0157447, Jul. 2016, doi: 10.1371/journal.pone.0157447.
23. C. Exley, and J. Kessler, "The Gender Gap in Self-Promotion," National Bureau of Economic Research Working Paper Series, vol. No. 26345, Oct. 2019, doi: 10.3386/w26345.
24. K. Kay, and C. Shipman, *The Confidence Code: The Science and Art of Self-Assurance – What Women Should Know*. New York: HarperCollins, 2014.
25. T. S. Mohr, "Why Women Don't Apply for Jobs Unless They're 100% Qualified," *Harvard Business Review*, Aug. 25, 2014. Accessed: Dec. 03, 2021. [Online]. Available: https://hbr.org/2014/08/why-women-dont-apply-for-jobs-unless-theyre-100-qualified.
26. J. Zenger, "The Confidence Gap in Men and Women: Why it Matters and How to Overcome It," *Forbes*, Apr. 08, 2018. Accessed: Dec. 03, 2021. [Online]. Available: https://www.forbes.com/sites/jackzenger/2018/04/08/the-confidence-gap-in-men-and-women-why-it-matters-and-how-to-overcome-it/?sh=455bb6cd3bfa.
27. L. P. Frankel, *Nice Girls Don't Get the Corner Office: 101 Unconscious Mistakes Women Make that Sabotage Their Careers*. New York: Warner Business Books, 2004. Accessed: Jan. 15, 2022. [Online]. Available: https://archive.org/details/nicegirlsdontget00fran.

3 It Begins with Pink and Blue

With Jessica L. Wanke

As discussed in Chapter 2, we all carry perceptions of the world based on our experiences, including expectations about appropriate roles and behaviors for each gender. These beliefs do not appear out of thin air but develop over time, beginning at birth. In this chapter, we will focus on how we develop our gender schema and its effect throughout our lives particularly as it relates to careers and professional development. While there are many theories regarding child development and gender identity development, the framework underlying this chapter draws from social cognitive theory in which conceptualizations of gender develop and change throughout an individual's life and are influenced by personal perceptions of self and broader sociocultural factors [1]. There are a number of studies and particularities related to gender development and we are unable to summarize them all here; however, we have included a few key points below.

TRUCKS OR DOLLS: DEVELOPMENT OF GENDER SCHEMA IN CHILDREN

Studies have found that children start to develop preferences for toys specific to gender (i.e., dolls versus trucks) at a young age, with one study specifically demonstrating preferences as young as 12.5 months [2]. In this study, male children were found to have more trucks than dolls at home while female children had about equal numbers of each. The process of developing gender-typed toy preferences was thought to be due to repeated exposure to specific toys (i.e., boys exposed to more trucks) and hypothesized to be implicit on the part of the parent – providing more dolls or trucks rather than overtly encouraging a child to play with a particular toy [2]. Importantly, this study demonstrates how parental influence is one of the earliest ways that children begin to develop their gender schema. While cultural shifts have resulted in many parents focusing on being gender-neutral, it is unknown whether this will result in long-term changes in gender schema, particularly as recent studies have shown persistence in gender-typing children's toys [3].

This study of toy preferences also reflects how we begin to develop interests and characteristics in response to social norms and expectations [2, 4]. While a large amount of research has been done over the past decades to determine the extent to which biology versus socialization shapes gender (i.e., nature versus nurture) [5, 6]; ultimately, it is likely a combination of both [1, 6]. Importantly though, when looking

DOI: 10.1201/9781003205814-4

at behavioral and physical characteristics, there are more differences and variability within each gender than between each gender [5, 7].

Much like the old saying "blue is for boys and pink is for girls," these conscious and unconscious messages reinforce binary expressions of masculinity and femininity rather than acknowledging that these traits exist on a continuum. While toy preference seems like a small example, it sets the stage for lifelong messages about appropriate play behaviors and ultimately appropriate life and work behaviors [2, 5]. The influence that parents, and later teachers and classmates, have on toy preferences and behaviors occur gradually over time after hundreds and hundreds of repeated interactions and exposures [2, 5].

MATH IS FOR BOYS: GENDER SCHEMA IN ELEMENTARY SCHOOL

As child development proceeds, the differences become more pronounced. These differences encompass many areas including perceptions about gender-appropriate interests and play, acceptable behavior, and belief in one's abilities [5, 8]. Studies have found significant differences in children's beliefs in their abilities beginning in elementary school. One study found that girls as young as six years old were less likely than boys to identify individuals of their own gender as being "really, really smart" [9]. That same study found that, as a result, they were less likely to choose to engage in activities that were said to be for "really, really smart" children [9]. This study indicates the gendered assumptions that girls have about their abilities and the long-term impacts on the activities they engage in, the classes they may choose, and ultimately what career they choose.

When it comes to Science, Technology, Engineering, and Math (STEM), girls tend to underestimate their abilities [10–12]. Further, they tend to believe that success is due to luck while failure is their fault [5, 8]. Boys and men do not share a similar perspective [5, 8]. This difference occurs despite any evidence that girls are inherently worse at math than boys [5, 7, 13]. These perspectives that girls have in elementary school and high school are a part of the gender schema that they develop throughout their lives in response to many implicit and explicit messages that they have received [5]. While girls have been shown to have more flexibility than boys in how they are able to express themselves along the spectrum of masculine to feminine during adolescence [14], their underlying conceptions of gender schema remain and they interact with others who carry similarly engrained gender schema [5]. Just like women in professional environments have to do more to be considered to be performing equally to men in similar jobs [15, 16], studies have found that girls are consistently rated lower by their teachers in mathematical proficiency than boys with comparable abilities [17, 18]. Further, there is a significant lack of girls and women portrayed in the media in STEM-related fields [19]. Numerous articles have been written about how seeing someone "like me" shapes the possibilities young children see for themselves [19, 20].

It is easy to see how these pieces add up over time. A ten-year-old girl who already believes that she isn't as good at intellectually challenging tasks, like math, and believes that if she fails, it is all her own fault, might have a single bad score on a math test which confirms her implicit belief that she is "bad at math." If she could consider other options, like that the teacher didn't teach the material well or the exam was just particularly challenging, she might focus on working harder and learning differently. However, if her teacher holds unconscious beliefs that female students are bad at math, this serves to reinforce the ten-year-old girl's belief about herself. These experiences accumulate and may shape her interest in continuing to take math or science classes, as well as her belief that she could pursue a career that requires math and science [5, 9].

THE ELEMENTARY PROBLEM

Learning – acquisition of skills, comprehension of concepts, application of strategies – is best achieved within real-life contexts. This is a common fact in education and as teachers, our question when approaching each lesson is: How do we make this relevant to our students? Advancing through the educational system, you may have even found yourself asking, "When will I ever use this? Why is this even important?" in a begrudging tone only teenagers can pull off. But the criticalness of this question has never been more apparent to me than as a new parent and current elementary school teacher.

Over the years, my students have come from a variety of backgrounds – socioeconomic status, ethnicity, culture, language, neurodiversities, academic abilities, and family make-ups. I always make it a personal mission to get to know my students and their strengths and interests both in and out of the classroom because their experiences in both can act as a bridge for learning. While building lessons upon prior knowledge developed in the classroom is critical, building lessons on common interests and background experiences out of the classroom can help peak students' motivation as well.

As a parent, this is where the reality of how we raise our little ones comes into play. When young children are asked to learn about advanced science concepts, like forces or plant growth, I can't help but wonder how much the background play experiences impact the relevance of the content and therefore how much the students grasp from the related science lessons in the classroom. I have to imagine that a student who has spent time playing in a dirt pile with toy excavators and dump trucks, building car tracks or marble runs, or exploring nature and planting a garden had a deeper personal connection to the science content than a student who has only seen science concepts in action on a screen. While it might be easy to see gender play a role in the reality of what toys and play experiences shape a person's early childhood, socioeconomic status and the opportunities it affords can play just as heavy of a role.

So here is the final question: How do we teach scientific concepts when background experiences play such a key role in connecting to the content but can vary so greatly?

Jessica Wanke – Elementary School Teacher

CAREER IMPLICATIONS

Choosing a Career

Both the beliefs that women develop about themselves and the beliefs that others (men and women) develop about them have significant long-term career implications [10, 21, 22]. Notably, unlike our gender schemas which put people in one bucket or another, both people and occupations are multidimensional [5]. Being a nurse requires strong problem-solving and analytical skills in addition to nurturing qualities. Being an engineer requires communication skills and the ability to work in a team in addition to mathematical and analytical abilities. However, within our gender schema, we have assigned one of these roles to be traditionally "female" and one of these to be traditionally "male." At the end of the day, these schema and our assumptions are incredibility limiting and restrictive for all individuals.

These limits affect more than just an individual's decision to pursue a particular career; they have long-term financial ramifications [16, 21, 23]. Further, they have broader societal ramifications as well. There is still a significant gender gap in innovation as established by current patent holders. At the current rate, it will take another 118 years to reach gender parity [24]. Just imagine all of the things that could have been invented and where we could be today if we had access to the ideas of all of the women who may have wanted to pursue careers in innovation but did not.

Succeeding in a Career

Women who pursue a career path that men traditionally dominate, like engineering, face another set of challenges driven by gender schema. As discussed, women are judged by a harsher metric – they are typically underrated at work while men are overrated [5, 23]. Each specific instance may not result in significant harm to a woman's career trajectory; however, these instances compound over time. Imagine during staff meetings that a male project manager regularly asks male engineers to look into technical issues while regularly asking the lone equally capable female engineer to take on responsibilities in the areas of planning or organizing the technical work. This may have many ramifications – the female engineer may doubt her abilities in design, the team members who already believed that the female engineer wasn't as competent may interpret this as confirmation bias that she is indeed not as competent, and the male engineer may receive further projects later on based on his experience in this project. Small instances build upon each other and widen the gap between the female engineer and her equally competent male colleagues.

This is similarly demonstrated in the field of academic medicine. While the number of female physicians is increasing, women are consistently less likely to be promoted to associate professor, full professor, or department chair [25]. These differences have persisted over the past 35 years. Notably, requirements for promotion frequently include giving invited talks, publications, and obtaining grants. You can see how the implicit biases that those reviewing grants may carry about the relative competency of women compared to men could have large ramifications on the career trajectory of women in academic medicine. Without receiving grants, it is harder for them to conduct research and therefore harder for them to publish. This ultimately makes it harder for women to get promoted.

Imposter Syndrome

Imposter syndrome describes the experience that individuals have of underestimating their abilities or thinking that they have not earned a particular success; it has been shown to affect both men and women [8, 26, 27]. Imposter syndrome includes frequent thoughts of "I'm not good enough" and feeling like someone else would do better. One can imagine all of the opportunities or ideas an individual may not pursue because of this underlying belief. While both genders experience imposter syndrome, for women working in environments where they are already viewed as inherently less competent and have to work harder to prove themselves, it can be much harder to overcome feelings of imposter syndrome. In addition to recognizing their own worth and abilities, they have to continue to prove their worth to those around them.

MOVING FORWARD

We know that we all carry with us gender schema and gender biases that shape how we make decisions and how we interact with others. While societal norms are changing, these gender schemas remain relatively fixed and result in cumulative disadvantages for women compared to men. This leads to the understandable question of what can be done. There have been many efforts to increase understanding of implicit biases and assumptions, increase mentorship, and increase role models for young girls, all of which should continue. However, given the systemic nature of the problem, it makes sense to consider system-based approaches.

Growth Mindset

One concept that has been used to encourage ongoing growth and learning is that of a growth mindset, a theory that has been developed and researched for decades [28]. A growth mindset is the belief that an individual can improve their abilities through things like working hard, assistance from others, and drawing upon useful strategies [28]. This is in contrast to an emphasis on an individual's abilities being entirely innate and therefore less amenable to growth through their efforts [28]. Growth-mindset interventions have been studied as a way to improve educational outcomes [28, 29]. A growth mindset approach focuses on encouraging students to persevere in the face of challenges, to actively seek help, and to be creative in their problem-solving approaches.

Given girls' inherent beliefs about luck contributing to success and their personal inabilities causing failures [5, 8], an environment that fosters a growth mindset seems ideal. Actively encouraging learning over outcomes shifts the focus to learning from mistakes, finding support from others, and persevering with the belief that one can improve. It further emphasizes that effort does not inherently imply a lack of ability, rather effort can be used to facilitate improvement in ability [28]. It can be useful at a young age when students are just developing their lifelong patterns of making and recovering from mistakes and can also be helpful in the workplace environment. A critical component of a growth-mindset environment, however, is that it is not the rewarding of effort without any focus on the outcome, rather it is fostering the belief that through effort and utilization of resources, improvement in outcomes is possible [28]. This freedom to make mistakes, and keep going, allows for innovation, collaboration, and, as the name suggests, growth [5]. Given this, in looking for a system-based approach to increase the number of women in engineering, it is important to find one that promotes an environment of growth mindset.

REAL-WORLD EXAMPLES

Let's look at some real examples that demonstrate how underlying gender schemas, which shape things like how men and women act in the workplace, impact our approach to increasing the retention of women engineers. Specifically, we can take an anecdote such as: at a women's technical conference, at what percentage of male attendance does the conference change from a women's conference into just another conference?

In this example, some might suggest that having men at a women's conference is helpful in that it brings in other points of view, creates advocates, and shows that men care. However, there is a clear tipping point at which the typical male-to-female interactions we see in male-dominated workplaces are apparent at the conference and may change the nature of the conference. We can assume that this would happen at a point before men are 50% of the conference attendees and, many years ago, a woman attendee at one of these events suggested that the rate is 30%. For the sake of this discussion, let's use that number as the tipping point at which male–female interaction patterns begin to dominate at the conference.

However, knowing that anything less than 50% of the participants being men still leaves women in the majority with the conference still structured around women's concerns, why is anything above that 30% tipping point but less than 50% an issue? A women's conference, or any conference for that matter, has two components: that which is controlled and that which is uncontrolled. The controlled elements include the presented topics, handling of break-out groups, and how these groups report back to the conference. The uncontrolled components might be limited to group discussion. During these group discussions, all of the normal male-dominated work environment elements may appear – variations in confidence levels, perceived or real power, men talking over women, and differences in experiences and knowledge. Even small differences in experiences and knowledge can have a significant impact. Assuming you are a good negotiator, you only need to know 5% more than the person you are negotiating against and you will win every time. Not 20%, but 5%. This potentially large impact of a small difference in knowledge explains why the dynamics of the personal interactions between men and women at a conference could completely flip in a scenario when men make up only 30% of the conference attendees despite the presenters and organizers establishing and controlling the agenda. While the information flow may be controlled, the dialogue surrounding that information is not. The combination of dialogue and personal interactions shapes the conference climate.

In another example of male–female interactions, a research study assessed the participation of male and female students in 13 university introductory biology classes in which more than 60% of the students were women. The authors found that "[w]hen students were asked to offer volunteer responses, 69% of classrooms showed a pattern of male-biased participation; across these classes, males on average spoke 63% of the time, even though they comprised 40% of the overall class" [30, p. 487]. This suggests that even when women are the majority, there will still be a gap in participation. In this case, men were more than twice as likely as women to answer the question even though they made up only 40% of the class. The study's authors note that the most logical approach to addressing this issue is for the instructor to randomly call on students. This is a change in structure and may be a method to build confidence in women students. We can see this proposed change as a change in the method of how the work is done, specifically establishing who will answer questions. Referring back to Figure 2.1, by calling on students and changing "how the work is done," the instructor gives a student an opportunity to build confidence, thereby creating a positive external influence.

Why do these examples matter? As we work through finding the root causes of our retention issue, we cannot just assume the problem will be solved if we create a 50–50 male-to-female ratio. Even at a 40–60 male-to-female split, the problem of unequal gender participation still existed. Therefore, when we use the words "male-dominated," we are not solely speaking of the percentage of men relative to women. Instead, we are speaking of situations in which men dominate the environment or situation for any ratio of men to women. From our examples above, we can assume that an environment may be male-dominated, with men making up only 30% to 40% of the group.

To fully address the challenges women face in engineering, we need to look for structural ways to change how the work is done in order to force changes in the dynamics of personal interactions while at the same time improving the overall performance of the firm or group.

THE ENGINEERING PIPELINE

Our engineering pipeline, Figure 3.1, shows the influence of gender schema and stereotypes as an individual progresses from birth to beginning Engineering School and then continuing on to their career. In Chapter 9, we will cover how an engineer gets their degree and how the structure of that learning mitigates some, or maybe a lot, of the bias, enabling some aspects of gender-neutral education. At the end of the pipeline, we show that the two areas of "Masculine" versus "Feminine" roles start out relatively equal at the beginning of a career but then diverge as time progresses. As we saw in our earlier story, in which the male manager regularly assigned the review of technical issues to male engineers while passing over the woman engineer, each small action can lead to an increasing divergence in career path. When

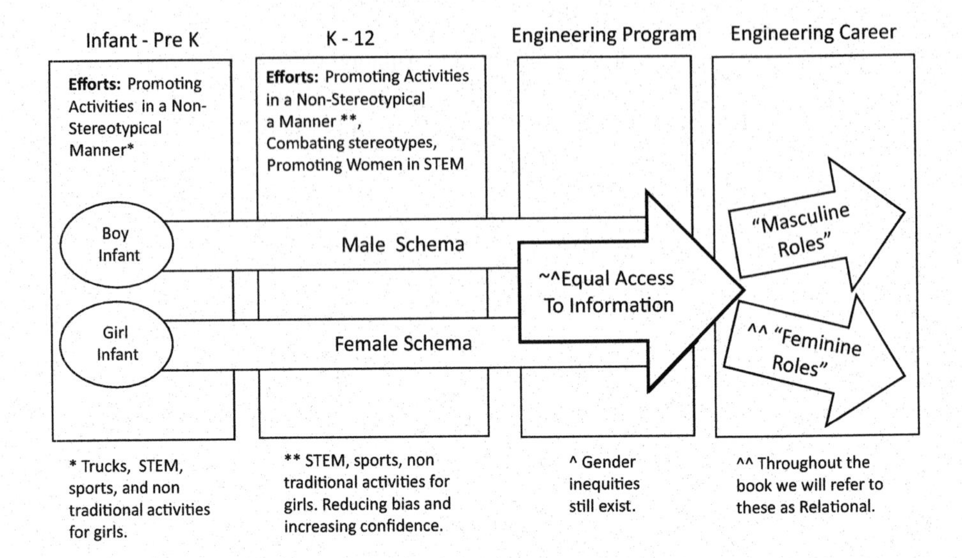

FIGURE 3.1 Engineering pipeline.

the woman engineer was a college hire, she may have been asked to look at technical issues at an equal rate compared to her male peers; however, over time her role shifted from a technical focus to a relational focus through no intentional actions of her own (the far right of Figure 3.1). In Chapter 5, we will further explore the relational roles that many women engineers migrate into as their career progresses as well as the long-term career impacts of a shift to relational roles.

Let's shift to looking at the **engineering** pipeline model from a pure process management perspective. Specifically, what does it take to double the number of female engineering graduates from the current rate of 22%?

1. With ~330 million people in the United States, there were roughly ~4.3 million 22-year-olds in 2019 [31].
2. That same year, there were roughly 146,000 bachelor engineering degrees awarded [32].
3. This means that in 2019 roughly 3% of the population of 22-year-olds (typical age of a college graduate) became engineers available for engineering work (146,000/4.3 million × 100 ≈ 3%).
4. With a goal of doubling the percentage of women graduating with engineering degrees from 22% to 44%, we are only talking about encouraging an additional 32,000 girls and women out of roughly 2 million to pursue an engineering degree or about 1.5%.

So, if we are looking at this from purely a process improvement effort, while a goal of doubling the output seems almost insurmountable, in reality we only need to recruit approximately 1.5% of the available input. However, we are not going to be able to increase our input if we don't address the retention issues, which is a major impediment to attracting girls and women to engineering. Therefore, the remainder of this book will focus on addressing the retention of women in engineering.

CONCLUSION

Beginning at an early age, we develop interests and beliefs about ourselves that are shaped by the broader societal and cultural contexts in which we live. These beliefs, such as beliefs about our mathematical abilities, can have long-term ramifications – including the career we choose and how successful we are in that career. While societal norms are changing, the underlying gender schemas are much slower to change. Further, the small disadvantages that women experience accumulate over time and contribute to gaps between men and women. Given the systemic nature of the problem, system-based approaches should be considered, specifically those that promote an environment of growth mindset. Importantly, increasing the number of women in engineering alone is not sufficient to overcome the current patterns of male–female communication that exist. Instead, the key is to find a structural way to change how the work is done that also changes the dynamics of personal interactions. We will now turn to using the field of medicine as a case study given its relative success at increasing gender diversity.

REFERENCES

1. K. Bussey, "Gender Identity Development," in *Handbook of Identity Theory and Research*, vol. 1 & 2, S. Schwartz, K. Luyckx, and V. Vignoles, Eds. Berlin, Germany: Springer, Springer Nature, 2011, pp. 603–628.
2. J. L. Boe, and R. J. Woods, "Parents' Influence on Infants' Gender-Typed Toy Preferences," *Sex Roles*, vol. 79, no. 5–6, pp. 358–373, Sep. 2018, doi: 10.1007/s11199-017-0858-4.
3. D. MacPhee, and S. Prendergast, "Room for Improvement: Girls' and Boys' Home Environments are Still Gendered," *Sex Roles*, vol. 80, no. 5–6, pp. 332–346, Mar. 2019, doi: 10.1007/s11199-018-0936-2.
4. N. Ellemers, "Gender Stereotypes," *Annual Review of Psychology*, vol. 69, pp. 275–298, Jan. 2018, doi: 10.1146/annurev-psych-122216-011719.
5. V. Valian, *Why so Slow? The Advancement of Women*. Cambridge: MIT Press, 1998.
6. A. H. Eagly, and W. Wood, "The Nature–Nurture Debates: 25 Years of Challenges in Understanding the Psychology of Gender," *Perspectives on Psychological Science*, vol. 8, no. 3, pp. 340–357, May 2013, doi: 10.1177/1745691613484767.
7. J. S. Hyde, "Gender Similarities and Differences," *Annual Review of Psychology*, vol. 65, no. 1, pp. 373–398, Jan. 2014, doi: 10.1146/annurev-psych-010213-115057.
8. K. Kay, and C. Shipman, *The Confidence Code: The Science and Art of Self-Assurance--What Women Should Know*. New York: HarperCollins, 2014.
9. L. Bian, S.-J. Leslie, and A. Cimpian, "Gender Stereotypes about Intellectual Ability Emerge Early and Influence Children's Interests," *Science*, vol. 355, no. 6323, pp. 389–391, Jan. 2017, doi: 10.1126/science.aah6524.
10. E. Seo, Y. Shen, and E. C. Alfaro, "Adolescents' Beliefs about Math Ability and Their Relations to STEM Career Attainment: Joint Consideration of Race/Ethnicity and

Gender," *Journal of Youth and Adolescence*, vol. 48, no. 2, pp. 306–325, Feb. 2019, doi: 10.1007/s10964-018-0911-9.
11. C. M. Ganley, and S. T. Lubienski, "Mathematics Confidence, Interest, and Performance: Examining Gender Patterns and Reciprocal Relations," *Learning and Individual Differences*, vol. 47, pp. 182–193, Apr. 2016, doi: 10.1016/j.lindif.2016.01.002.
12. S. Nix, L. Perez-Felkner, and K. Thomas, "Perceived Mathematical Ability Under Challenge: A Longitudinal Perspective on Sex Segregation among STEM Degree Fields," *Frontiers in Psychology*, vol. 6, pp. 530, Jun. 2015, doi: 10.3389/fpsyg.2015.00530.
13. S. M. Lindberg, J. S. Hyde, J. L. Petersen, and M. C. Linn, "New Trends in Gender and Mathematics Performance: A Meta-Analysis," *Psychological Bulletin*, vol. 136, no. 6, pp. 1123–1135, 2010, doi: 10.1037/a0021276.
14. M. L. Halim, D. N. Ruble, and D. M. Amodio, "From Pink Frilly Dresses to 'One of the Boys': A Social-Cognitive Analysis of Gender Identity Development and Gender Bias: Changes in Gender Identity and Gender Bias," *Social and Personality Psychology Compass*, vol. 5, no. 11, pp. 933–949, Nov. 2011, doi: 10.1111/j.1751-9004.2011.00399.x.
15. E. H. Gorman, and J. A. Kmec, "We (Have to) Try Harder: Gender and Required Work Effort in Britain and the United States," *Gender & Society*, vol. 21, no. 6, pp. 828–856, Dec. 2007, doi: 10.1177/0891243207309900.
16. A. Joshi, J. Son, and H. Roh, "When Can Women Close the Gap? A Meta-Analytic Test of Sex Differences in Performance and Rewards," *Academy of Management Journal*, vol. 58, no. 5, pp. 1516–1545, Oct. 2015, doi: 10.5465/amj.2013.0721.
17. J. Tiedemann, "Gender-Related Beliefs of Teachers in Elementary School Mathematics," *Educational Studies in Mathematics*, vol. 41, no. 2, pp. 191–207, 2000, doi: 10.1023/A:1003953801526.
18. J. R. Cimpian, S. T. Lubienski, J. D. Timmer, M. B. Makowski, and E. K. Miller, "Have Gender Gaps in Math Closed? Achievement, Teacher Perceptions, and Learning Behaviors Across Two ECLS-K Cohorts," *AERA Open*, vol. 2, no. 4, pp. 233285841667361, Oct. 2016, doi: 10.1177/2332858416673617.
19. The Lyda Hill Foundation & The Geena Davis Institute on Gender in Media, "Portray Her: Representations of Women STEM Characters in Media," The Lyda Hill Foundation & The Geena Davis Institute on Gender in Media, Online, c 2022. Accessed: Jan. 15, 2021. [Online]. Available: https://seejane.org/wp-content/uploads/portray-her-full-report.pdf.
20. S. Varalli, "Seeing is Believing: The Importance of Visible Role Models in Gender Equality," *The Art Of*, May 09, 2017. Accessed: Jan. 16, 2022. [Online]. Available: https://www.theartof.com/articles/seeing-is-believing-the-importance-of-visible-role-models-in-gender-equality.
21. C. Exley, and J. Kessler, "The Gender Gap in Self-Promotion," National Bureau of Economic Research Working Paper Series, vol. No. 26345, Oct. 2019, doi: 10.3386/w26345.
22. L. N. Mackenzie, J. Wehner, and S. J. Correll, "Why Most Performance Evaluations are Biased, and How to Fix Them," *Harvard Business Review*, Jan. 11, 2019. Accessed: Jan. 15, 2021. [Online]. Available: https://hbr.org/2019/01/why-most-performance-evaluations-are-biased-and-how-to-fix-them.
23. J. F. Madden, "Performance-Support Bias and the Gender Pay Gap among Stockbrokers," *Gender & Society*, vol. 26, no. 3, pp. 488–518, Jun. 2012, doi: 10.1177/0891243212438546.
24. A. Bell, R. Chetty, X. Jaravel, N. Petkova, and J. Van Reenen, "Who Becomes an Inventor in America? The Importance of Exposure to Innovation," *Quarterly Journal of Economics*, vol. 134, no. 2, pp. 647–713, May 2019, doi: 10.1093/qje/qjy028.

25. K. P. Richter *et al.*, "Women Physicians and Promotion in Academic Medicine," *The New England Journal of Medicine*, vol. 383, no. 22, pp. 2148–2157, Nov. 2020, doi: 10.1056/NEJMsa1916935.
26. R. L. Badawy, B. A. Gazdag, J. R. Bentley, and R. L. Brouer, "Are All Impostors Created Equal? Exploring Gender Differences in the Impostor Phenomenon-Performance Link," *Personality and Individual Differences*, vol. 131, pp. 156–163, Sep. 2018, doi: 10.1016/j.paid.2018.04.044.
27. A. R. Vaughn, G. Taasoobshirazi, and M. L. Johnson, "Impostor Phenomenon and Motivation: Women in Higher Education," *Studies in Higher Education*, vol. 45, no. 4, pp. 780–795, Apr. 2020, doi: 10.1080/03075079.2019.1568976.
28. C. S. Dweck, and D. S. Yeager, "Mindsets: A View From Two Eras," *Perspectives on Psychological Science*, vol. 14, no. 3, pp. 481–496, May 2019, doi: 10.1177/1745691618804166.
29. D. S. Yeager *et al.*, "A National Experiment Reveals Where a Growth Mindset Improves Achievement," *Nature*, vol. 573, no. 7774, pp. 364–369, Sep. 2019, doi: 10.1038/s41586-019-1466-y.
30. S. L. Eddy, S. E. Brownell, and M. P. Wenderoth, "Gender Gaps in Achievement and Participation in Multiple Introductory Biology Classrooms," *CBE: Life Sciences Education*, vol. 13, no. 3, pp. 478–492, Sep. 2014, doi: 10.1187/cbe.13-10-0204.
31. United States Census Bureau, "U.S. and the World Population Clock," *U.S. and World Population Clock*. https://www.census.gov/popclock/.
32. American Society for Engineering Education, "Engineering and Engineering Technology by the Numbers 2019," American Society for Engineering Education, Washington, DC, Report, 2020. Accessed: Dec. 03, 2021. [Online]. Available: https://ira.asee.org/wp-content/uploads/2021/06/Engineering-by-the-Numbers-2019-JUNE-2021.pdf.

Part I

Summary

Throughout Part I, we focused on laying the groundwork of the overall problem.

- In Chapter 1, we established that two decades of focused effort to retain women in engineering delivered marginal progress. However, medicine has had a dramatically different result, and women now make up more than 50% of students graduating from medical schools.
- In Chapter 2, we described gender schema, implicit bias, and confidence, and how these interact and impact retention of women in engineering. Figure 2.1 provided a simple model to help us understand how these complex elements are interrelated. Most significantly though, using this model, we introduced the element of "how the work is done" and how it can directly impact both confidence and bias.
- In Chapter 3, we focused on the development of gender schema and how girls develop beliefs about their abilities in Science, Technology, Engineering, and Math (STEM) at a young age, which have long-term impacts on their career choices. Equal access to knowledge through an engineering program might mitigate some of the effects of individual's inherent beliefs (see Figure 3.1); however, there are no such mitigating factors in the workforce.

As we move into Part II, we will examine medicine as a model for engineering, emphasizing job satisfaction as this is crucial to employee retention. We will take a detailed look at all of the areas impacting women in engineering and establish what we believe are the root causes of our retention issue.

DOI: 10.1201/9781003205814-5

Part II

Analyzing the Problem

Using an Engineering Process

4 Medicine as a Model for Engineering

Creating Job Satisfaction

On the face of it, a physician's day-to-day work looks very different from the day-to-day work of an engineer. However, if we shift away from the specific work activities of physicians and engineers and focus instead on the skills and abilities required for each field, we find a striking number of similarities. Specifically, both engineers and physicians use critical thinking, complex problem solving, judgment and decision-making, inductive and deductive reasoning, and active listening [1]. Both fields require a base of fundamental knowledge paired with critical thinking to make decisions in challenging situations. The development of these core skills allows physicians and engineers to have a high level of work variation, performing various work activities during their career and even within one job. The same mechanical engineering degree can allow an engineer to work on an oil rig or on a product development team creating the next innovation that may completely change the world. The same Internal Medicine physician can manage blood pressure in the outpatient setting or treat complicated infections in hospitalized patients.

Why does this comparison matter? At the end of the day, physicians will move toward 50% women while engineering is starting from 15% women with minimal movement [2, 3]. It is important to note, though, that while the growth made in the sheer number of women in medicine over the past 50 years is substantial, there remain a number of disparities between women and men in medicine including salary, promotion, leadership positions, and medical specialty choice [4, 5]. Examining gender distribution in medicine between medical specialties highlights one example of this disparity as women made up roughly 83% of Obstetrics/Gynecology residents and men made up about 85% of Orthopedic Surgery residents in 2019 [5]. Importantly, though, the percentage of female residents within male-dominated fields has increased over time [6]. In engineering, similar challenges exist for women in terms of differences in career paths, lower median pay, concerns about opportunities for promotion, and choice of engineering discipline [7–10]. For example, within the field of engineering, women earn 51.7% of Environmental Engineering Bachelor's degrees but only 15.7% of Mechanical Engineering Bachelor's degrees [8]. While the example of the Environmental Engineering degree is notable in that it has a high level of gender parity, it is also only a small percentage (less than 2%) of overall engineering degrees [8]. Despite the ongoing challenges medicine faces in regard to gender equity, the success that the field has had in reducing gender disparity compared to the field of engineering in combination with the similarities in skills and abilities required of physicians and

DOI: 10.1201/9781003205814-7

engineers makes medicine a potential model for the future direction of engineering. By understanding the factors contributing to medicine's success, we can find new methods to increase the retention of women in engineering.

INCREASING GENDER INTEGRATION: MEDICINE VERSUS ENGINEERING

As career opportunities continue to open for women, there are some jobs that may simply have more appeal than others. The difference in appeal may be related to the nature of the work itself. Some studies have found that when choosing a career, women prioritized altruism and were interested in fields that were more people-oriented while men prioritized status and were interested in fields that were things-oriented [11, 12]. These internal factors may account for a portion, although not all, of the difference in gender disparity within medicine and engineering particularly as studies have also found that differences in interest may not drive career choice [11, 13]. Medicine is generally understood as an altruistic career, focusing on helping people. Engineering, with possibly the exception of Environmental Engineering, does not always carry a similar perception [14]. While these internal factors are not modifiable in the sense that they are value-guided decisions made by individuals with autonomous choices, the perception of engineering as a career aligned with altruism is likely modifiable [14]. Further, these value preferences may be related to our underlying gender schema, specifically preferences for specific values and interests may be based on underlying gender norms, and therefore may change over time. Notably, though, there are many external factors that have also been found to contribute to why women choose one career path over the other including work culture, level of support available, feelings of belonging, and level of gender integration, which are modifiable and are therefore targets for intervention [9].

Historically, Science, Technology, Engineering, and Math (STEM)-related organizations have focused on the following five areas to attract and retain women engineers*:

1. Corporate-wide initiatives in schools to provide educational STEM experiences for girls
2. Efforts to build an inclusive and positive work environment
3. Efforts to address bias within the workplace
4. Ensuring women have an opportunity to move into leadership positions
5. Increasing role models and mentorship

The field of medicine has similarly focused on these areas through organizations designed to support women on both institutional levels and national levels (i.e. American Medical Women's Association, Association of Women Surgeons) and

* These five areas summarize a broad spectrum of activities, driven by the first author's observations and experiences. Examples of programs can be found on the Society of Women Engineers (SWE) website [15].

initiatives targeted at reducing bias within the workplace, improving mentorship, and increasing the opportunity and ability for women to move into leadership positions [16–19]. Assuming both engineering and medical organizations are roughly similar in their ability to implement the above initiatives, there must be other factors contributing to medicine's relative success in reducing gender disparities compared to the field of engineering. One notable difference between the careers is the method of knowledge acquisition and the development of necessary skills.

The vast majority of new physicians, specifically defined as individuals who have just graduated from medical school, enter into a three-to-seven-year residency program. During their time in residency, they provide clinical care, gain knowledge related to their specific specialty, and develop the core skills needed to practice the art of medicine – the ability to apply their scientific knowledge to the individual sitting in front of them. In contrast, new engineers, defined specifically as those who have just graduated from a bachelor's or master's program with a degree in engineering, enter the workforce with the subsequent training being much more variable. A better understanding of the impact that this difference in training has on women's experience in the workplace may result in a way to target the system of engineering rather than only targeting the individuals working within it.

UNDERSTANDING THE DIFFERENCE: THE KNOWLEDGE AVAILABILITY GAP

There is a significant difference between the knowledge readily available to physicians compared to the knowledge readily available to engineers. We can create a rough estimate of the web-based knowledge available to an engineer or physician by asking one simple question: What percent of your knowledge,* meaning foundational facts and information rather than the acquisition of skills like critical thinking or technical skills, are you able to find on the web, either internal or external to your organization? When asking this question of a physician, anecdotal data suggest that 70%† of the information necessary to answer questions that arise during day-to-day work is web-based – including medical journals and textbooks that are available online. Of the remaining 30%† of their learning, about half might come directly from the patient with the remaining 15%† of their necessary knowledge coming from a peer or supervising physician.† While the majority of physicians' clinical reasoning, decision-making skills, and application of knowledge are obtained through clinical experience and learning from other clinicians, the underlying knowledge base can be obtained through external sources. When you ask the same question of an engineer doing detail-oriented engineering work, anecdotal data suggest that only 20%

* Of note, here we are distinguishing knowledge from skills acquisition. Specifically, knowledge being the foundational facts and information and skills acquisition being the development of critical thinking skills, ability to apply knowledge to particular situations, and technical skills. Clinical reasoning and decision-making fall under the broad category of skills acquisition in this model. Learning is then the combination of knowledge and skills acquisition.

† These percentages are rough estimates and vary significantly by specialty, particularly surgical specialities.

to 30% of their knowledge for day-to-day work is web based. This means that an engineer would be dependent on peers and managers or possibly trial and error for up to 80% of their knowledge.

A quick internet test gives an eye-opening view into this reality at a basic level. There is an immense amount of medical knowledge available on the web to virtually anyone with a computer. Good or bad, many individuals have experienced looking up information on medical symptoms or a diagnosis and using what feels like a doctor on the internet. The information available on the pages and pages of websites can be overwhelming. When added to the medical sites that are only accessible by those in the medical field, the amount of accessible knowledge increases significantly. While the wealth of knowledge might be overwhelming to non-physicians, physicians have been trained to utilize it effectively and recognize its limitations. In contrast, if you chose an engineering-related topic or your latest computer problem and searched for technically relevant information, you might struggle to find even a few areas of help. Engineers face a significant knowledge availability gap compared to physicians.

An assumed root cause of the knowledge availability gap for engineers may be a lack of investment by engineering firms in methods to build knowledge databases. Collectively, engineering firms may not focus on or be motivated to create knowledge bases within their organizations. While they are often worried about competitors gaining access to sensitive information, it is also challenging to generalize the knowledge. There may be hundreds of physicians in medicine that can benefit from specific medical information. In contrast, only a handful of engineers may benefit from specific areas of technical knowledge, and the group of engineers most likely to benefit may all be internal to an organization and potentially even working on the same team. The difficulties with access to knowledge are compounded by the fact that engineering teams may be isolated from other teams even within the same organization. This makes the ability of engineers to glean information from teams outside of their own nearly impossible and the ability of an individual to learn from their own team that much more critical.

Ultimately, the difference in the level of dependency on peers or managers for learning has a profound impact on how the work is done and how an individual learns. During training in residency, physicians are able to use their time at work to focus on learning skills rather than on knowledge acquisition. From supervising physicians and other clinical team members, they learn the art of medicine and practice applying knowledge to different clinical situations. For example, when a patient comes to the hospital with chest pain, there is information readily available about tests you can order and diagnoses you should consider. There are also multiple decision tools that can help clinicians decide the risk and likelihood of a particular cause of the patient's chest pain. However, patients do not present to the hospital precisely as they do in textbooks. Instead of needing to ask their supervising physician basic questions about evaluation and treatment, a resident physician can focus on asking questions that can't be easily learned online – like how to apply the available knowledge to this particular patient presenting at this particular time. The next time the new resident evaluates a patient with chest pain, they have a clinical reasoning framework to use – one that they can continue to develop during training and throughout their career.

The training structure in place during residency allows for the gradual development of autonomy and confidence over time while emphasizing the learning of critical thinking and decision-making skills. In engineering, a significant amount of time on the job is spent trying to acquire the necessary knowledge. This means that not only is less time spent learning other important skills, but also less time is spent building confidence, creating a sense of belonging, and creating value. The availability of knowledge gap becomes a greater issue for women because of their typical marginalized status within the organization.

JOB SATISFACTION

The issue of dependency on others for knowledge goes beyond just creating job-related learning challenges for engineers, it also affects key elements related to retention. To fully understand why this is the case, we first have to examine the factors related to job retention. We propose three general factors that influence whether an individual stays or leaves a job: (1) work climate, (2) satisfying work, and (3) control over career and career trajectory.

WORK CLIMATE

As we have discussed previously, work climate plays a role in both career choice and retention of female engineers. The components of a positive work climate include being valued and respected, work–life balance, and having available role models [9]. The nature of the work climate is driven by a number of factors including implicit bias, the general environment of the organization, and whether it is a "progressive firm." A positive work climate contributes to increased satisfaction at work and therefore increased retention.

SATISFYING WORK

To understand the meaning of satisfying work, we can draw on the work of Malcolm Gladwell in his book *Outliers*. In the book, he writes: "Those three things -- autonomy, complexity, and a connection between effort and reward -- are, most people will agree, the three qualities that work has to have if it is to be satisfying" [20, p. 149].

THE COMPONENTS OF SATISFYING WORK

1. *Autonomy* – being your own boss, being responsible for decisions, and being able to make them.
2. *Complexity* – complex problems, increasing levels of complexity allow for further learning.
3. *Relationship between effort and reward* – seeing a positive end result of your effort.

Based on these three components, we suggest that the knowledge availability gap results in overall decreased satisfaction with work. A lack of access to reusable knowledge results in a higher level of dependence on others, leading to a lack of autonomy. This lack of autonomy reduces the complexity of work an individual can do and their lack of independence causes an inability to separate the complex from the trivial, resulting in someone becoming bogged down by both. In the worst situation, an engineer may spend time working through what they see as a complex problem only to find out that many other people have previously solved it. In a scenario classically seen in engineering organizations, an early-in-career engineer presents their solution to others only to be told by a senior engineer, "Good job solving it, but you should have just talked to me. I could have given you the answer. It's right here on a slide deck." Without access to basic knowledge, the early-in-career engineer was unable to determine which problems they should focus their efforts on. Further, they could not determine the best set of questions to ask and when to ask them. In the context of our gender schema, and women being viewed as less competent than equally competent men, a woman engineer making the decision to ask questions or seek out help has the added challenge of considering the potential consequences of being viewed negatively.

Ultimately, being dependent on others for learning impairs an individual's ability to gradually develop autonomy, learn how to solve increasingly complex problems, and see a clear correlation between their effort and the value others find in their work. These factors contribute to a lack of meaning and purpose in work resulting in decreased retention.

Control over Career

Control over career is the ability to establish what you want to learn or do, being enabled to do it, and having the opportunity to achieve it. Each of these requires deeper-level work to actually carry out; however, they all start with having access to the knowledge required to succeed in the effort. We see this when we research a job that we are interviewing for. In order to know if this is the type of job we would like and could excel at, we need access to information about what the job is, how the work is done, and how it fits with our personal goals. Through the interview process, we need to articulate what capabilities we bring to the job and how it fits with our interests. While few of us regularly prepare for or participate in job interviews, this aspect of control over career plays out in our daily work environment. For example, when you find a brand-new opportunity within your current job, you need to be able to understand it using the information you have access to and then, after determining if it meets your goals, pursue it.

In moving a career in the desired direction, dependency on others for learning is a significant inhibitor. It limits one's ability to understand what might be possible and impedes someone from actually working toward their goals and creating their own opportunities. An individual's ability to independently acquire the necessary knowledge is critical to control over career.

COMPONENTS OF CONTROL OVER CAREER

1. *Establishing the direction of the desired work* – being able to determine what you want to learn or do while having direct visibility of the opportunities in front of you.
2. *Being enabled to do the work* – being personally enabled to have a sense of control over the work and the effort required; having direct access to the information and knowledge required to achieve a goal that drives personal growth.
3. *Having the opportunity* – being able to create your own opportunity.

Let's look at a real example of integrating satisfying work, control over career, and confidence.

CREATING SATISFYING WORK WHILE BUILDING CONFIDENCE

It was early on a Saturday morning when I (Bob) came down the stairs to see Alissa sitting at our kitchen table using her computer. Alissa, home on vacation, was in her second year of residency and managing a heavy workload. I casually asked if she was doing work and she replied, "Yes." She was getting ready to start a new rotation in the Medical Intensive Care Unit (MICU) on Monday, and as the resident physician, she would be responsible for eight of the intensive care unit (ICU) beds. Curious, I asked if she was reviewing the patients that she would be responsible for. She explained that she had no idea who she would be treating at this time because she didn't know who would be discharged or admitted between now and then. Instead, she was preparing for the rotation by reviewing the equipment she would be using to best manage her patients' care.

At our kitchen table, on the other side of the country from the hospital in which she worked, Alissa was working through what she needed to know in order for her to do her job. After being highly trained for a specific area of care, she had the tools and information readily available for her to review. From this, Alissa was able to ensure she could provide proper care for her patients, with confidence, while being in control of her new rotation because of her access to critical pieces of knowledge. I was amazed at the level of control she personally had over her career.

Unfortunately, this level of available knowledge and career control is not typically available in most engineering environments. Without adequate control over their career, individuals are much more likely to leave a job.

TRAINING APPROACH: MEDICINE VERSUS ENGINEERING

As we look at the training approach of early physicians versus early engineers, we start with the basic understanding that the amount of knowledge that a physician in residency has access to is vast in comparison to that of an engineer new to their career. Available knowledge provides the foundation for further training in residency and is a key equalizing factor between resident physicians. Rather than trying to find knowledge, resident physicians can focus their questions and learning objectives at work on clinical decision-making and communication skills that fall outside of the facts available in the knowledge database.

A physician in residency typically works in a "teaching hospital" with an identified focus on meeting the health needs of patients while training health professionals. Resident physicians are directly working with and supervising medical school students as well as resident physicians who are earlier in their training. This inherently leads to an organization with a culture of learning. The training system is designed to advance each physician from a learner to an educator as they rotate through various areas of clinical medicine, gaining gradual autonomy and independence.

The combined access to an immense amount of knowledge and a learning environment helps to create a more level playing field between men and women in medicine by putting the female physician in control of her learning. This is further amplified by a curriculum put in place by a "management" team, which includes a variety of specialty and practice-type rotations as well as regular lectures for personal and professional development. The end goal is the completion of residency having met nationally standardized program requirements in order to be eligible to pass board certifications.

Unlike the development of a physician, there is little that is consistent from one engineer's experience to another across firms, or sometimes even within very large firms, as it relates to their training when they enter their first job. Some organizations may have a structured employee onboarding process that could include some aspect of technical training; however, just as likely, it may focus on primarily administrative activities or not exist at all.

Ultimately, the most basic step for establishing the effectiveness and efficiency of training is as simple as understanding someone's rate of learning and control over learning. Both engineers and physicians are reliant on their ability to learn new areas of knowledge quickly and then apply them competently. The faster that someone learns, the more productive they are and consequently the more productive the organization as a whole. In addition, the faster someone learns, the more opportunities they will have for personal growth and the more satisfying the work will be. The implications of this relationship are simple but critical: when employees have ready access to available knowledge, they will learn faster, they will be more productive, and thrive. Thriving employees means a thriving company. Alternatively, when employees do not have ready access to information, they learn slower, which means the organization learns slower.

THE WORK OF ENGINEERING

When narrowed down to its singular core focus, medicine's emphasis is on solving problems related to the health of the human body. Like medicine, engineering's fundamental goal is to identify and solve problems that matter to people, the key difference is the breadth of these problems. Engineering is immensely diverse in its work and business goals. An engineer's value add within product development is to create usable knowledge that enables profit and revenue for the firm. The work varies across a wide spectrum and can be managed with a high degree of independence that is unmatched by most other professions. Although most engineering industries have external regulatory requirements, such as safety, engineering as a whole has significant freedom in the way that innovation occurs. The question is how long does it take to create that innovation and deliver it to a customer. At its core, though, every engineering firm has the same ultimate goal: to generate profit and revenue through the delivery of a product or service. This goal, regardless of how the work gets done, is generally internally driven by the engineering firm or externally by a customer.

Ultimately, this diverse set of work, diverse set of goals for each firm, and diverse set of risks and rewards make engineering very different from medicine. However, success remains completely dependent on the rate of learning. Although the work of engineering currently may not be centered on learning or focused on accessible knowledge, that does not mean that it shouldn't be or can't be. A structured learning method and access to available knowledge are key to increasing the learning rate and advancement of employee careers and ultimately to increasing retention of women in engineering.

CONCLUSION

In comparing engineering to medicine, we find a stark difference in the success of achieving gender balance both in the attraction of women to the profession, as measured by the gender make-up in schools, and in retention as shown in statistical surveys. Each of these professions has its unique challenges, but medicine has achieved dramatically different results in attracting and retaining women in the field. The unique difference between training in engineering and medicine is that medicine has a structured training process that has created an infrastructure to facilitate learning. Drawing upon the training process of medicine provides a case study for training in engineering.

REFERENCES

1. O*NET OnLine, "Browse STEM Occupations," *O*NET OnLine*, 2018. https://www.onetonline.org/find/stem?t=0 (accessed Jul. 31, 2021).
2. R. Fry, B. Kennedy, and C. Funk, "STEM Jobs See Uneven Progress in Increasing Gender, Racial and Ethnic Diversity," Pew Research Center, Apr. 2021. [Online].

Available: https://www.pewresearch.org/science/wp-content/uploads/sites/16/2021/03/PS_2021.04.01_diversity-in-STEM_REPORT.pdf.
3. P. Boyle, "Nation's physician workforce evolves: more women, a bit older, and toward different specialties," *AAMCNews*, Feb. 02, 2021. Accessed: Jan. 15, 2022. [Online]. Available: https://www.aamc.org/news-insights/nation-s-physician-workforce-evolves-more-women-bit-older-and-toward-different-specialties.
4. N. B. Lyons *et al.*, "Gender Disparity Among American Medicine and Surgery Physicians: A Systematic Review," *The American Journal of the Medical Sciences*, vol. 361, no. 2, pp. 151–168, Feb. 2021, doi: 10.1016/j.amjms.2020.10.017.
5. B. Murphy, "These Medical Specialties Have the Biggest Gender Imbalances," *American Medical Association (AMA)*, Oct. 01, 2019. https://www.ama-assn.org/residents-students/specialty-profiles/these-medical-specialties-have-biggest-gender-imbalances (accessed Dec. 05, 2021).
6. C. C. Chambers, S. B. Ihnow, E. J. Monroe, and L. I. Suleiman, "Women in Orthopaedic Surgery: Population Trends in Trainees and Practicing Surgeons," *The Journal of Bone and Joint Surgery*, vol. 100, no. 17, pp. e116 (1–7), Sep. 2018, doi: 10.2106/JBJS.17.01291.
7. M. T. Cardador, and P. L. Hill, "Career Paths in Engineering Firms: Gendered Patterns and Implications," *Journal of Career Assessment*, vol. 26, no. 1, pp. 95–110, Feb. 2018, doi: 10.1177/1069072716679987.
8. American Society for Engineering Education, "Engineering and Engineering Technology by the Numbers 2019," American Society for Engineering Education, Washington, DC, Report, 2020. Accessed: Dec. 03, 2021. [Online]. Available: https://ira.asee.org/wp-content/uploads/2021/06/Engineering-by-the-Numbers-2019-JUNE-2021.pdf.
9. N. A. Fouad, and R. Singh, "Stemming the Tide: Why Women Leave Engineering," Center for the Study of the Workplace at University of Wisconsin - Milwaukee, 2011. [Online]. Available: https://www.energy.gov/sites/prod/files/NSF_Stemming%20the%20Tide%20Why%20Women%20Leave%20Engineering.pdf.
10. Society of Women Engineers (SWE), "Earnings Gap," Society of Women Engineers (SWE), 2021. https://swe.org/research/2016/earning-gap/ (accessed Feb. 02, 2022).
11. J. M. Lakin, V. A. Davis, and E. W. Davis, "Predicting Intent to Persist from Career Values and Alignment for Women and Underrepresented Minority Students," *International Journal of Engineering Education*, vol. 35, no. 1(A), pp. 168–181, 2019.
12. R. Su, J. Rounds, and P. I. Armstrong, "Men and Things, Women and People: A Meta-Analysis of Sex Differences in Interests," *Psychological Bulletin*, vol. 135, no. 6, pp. 859–884, Nov. 2009, doi: 10.1037/a0017364.
13. B. Ertl, and F. G. Hartmann, "The Interest Profiles and Interest Congruence of Male and Female Students in STEM and Non-STEM Fields," *Frontiers in Psychology*, vol. 10, pp. 897, Apr. 2019, doi: 10.3389/fpsyg.2019.00897.
14. J. M. Lakin, D. Marghitu, V. Davis, and E. Davis, "Introducing Engineering as an Altruistic STEM Career," *The Science Teacher*, vol. 88, no. 4, Apr. 2021, Accessed: Dec. 06, 2021. [Online]. Available: https://www.nsta.org/science-teacher/science-teacher-marchapril-2021/introducing-engineering-altruistic-stem-career.
15. Society of Women Engineers, "SWE Programs and Resources," Society of Women Engineers (SWE), 2021. https://swe.org/research/2021/swe-programs-and-resources/ (accessed Feb. 02, 2022).
16. AAMC, "Gender Equity in Academic Medicine," AAMC, 2022. https://www.aamc.org/news-insights/gender-equity-academic-medicine (accessed Feb. 24, 2022).
17. J. L. Welch, H. L. Jimenez, J. Walthall, and S. E. Allen, "The Women in Emergency Medicine Mentoring Program: An Innovative Approach to Mentoring," *Journal of Graduate Medical Education*, vol. 4, no. 3, pp. 362–366, Sep. 2012, doi: 10.4300/JGME-D-11-00267.1.

18. American Medical Women's Association (AMWA), "AMWA Invests in You with Quality Mentorship," American Medical Women's Association (AMWA), 2022. https://www.amwa-doc.org/about-amwa/member-benefits-amwa/mentoring/ (accessed Jan. 16, 2022).
19. L. P. Fried *et al.*, "Career development for women in academic medicine: Multiple interventions in a department of medicine," *JAMA*, vol. 276, no. 11, pp. 898–905, Sep. 1996.
20. M. Gladwell, *Outliers*. New York: Little, Brown and Company, 2008.

5 Through the Lens of an Engineer

It's early Monday morning and you are standing at the front of a room before half a dozen of the most capable people in the organization. A wall of blank whiteboards is behind you, freshly cleaned over the weekend. You have a marker in one hand and a piece of paper in the other. People are scattered throughout the room, sitting on the tabletops, and ready to work. No one brought a laptop. No one is on their phone. No one walks in late. From your Friday e-mail, they understand the severity and importance of the situation; this understanding will help them be successful. However, you know that the success of the team's work will largely be driven by its diversity. The variety of problem-solving approaches within the team and the team members' range of experiences and knowledge. Last week you were given the problem of addressing the unacceptable performance of a process. A process whose results have lagged behind the rest of the industry for years. Meanwhile, a competitor's process is attracting top talent and delivering year-over-year gains at your organization's expense. As your process fails to deliver the required results, its lackluster performance bleeds over to impact virtually every other part of the business. The revenue is not what it could be, and the delivery of innovation is in no better shape. The process struggles to meet "time to market" commitments and is challenged to meet key customer needs. Meanwhile, some of your most capable people, people who the organization has been training for years, see better opportunities outside of the company and take them.

If this story was about a business leader responsible for a struggling organization with an uncompetitive product line, it could be a preamble to a business novel – but it isn't. In this story, you are an engineering leader at a technical firm. The urgent issue facing you is your firm's inability to attract and retain women engineers with your competitor being none other than the medical industry. While the medical field is not an industry competitor, it is a direct workforce competitor as a Science, Technology, Engineering, and Math (STEM)-oriented career path. The medical industry has proven its ability to attract top talent at a higher rate than your industry; in this case, women leaving high school with an interest in STEM careers and then choosing a career in medicine over engineering. To make matters worse, at your organization, the area you have direct control over, many of the capable women who did decide to pursue engineering are leaving for better opportunities. In many cases, they are leaving the field altogether. These are engineers who you have invested time and energy into training and, in some cases, women who you personally hired. Despite your best efforts to recruit and retain women engineers, they make up less than 20% of your engineering staff. This lack of gender diversity is contributing to your business challenges – decreasing innovation and inhibiting revenue growth. Your motivation to

DOI: 10.1201/9781003205814-8

improve this scenario is simple: create a diverse work environment in order to deliver better business results. The only thing saving you is that your true industry competitors, those delivering competing products or services, are dealing with the same lack of diversity. In the end, everyone within your industry, including your organization, is average. However, the opportunity to outpace the rest of the industry is clearly there and you have been tasked with taking advantage of it.

THE PROBLEM

Let's play out the story above a little further. Standing at the front of the room, you first make sure your team is clear on the current performance of the engineering system. You begin by listing the simple process measures on the farthest left whiteboard.

THE PROCESS MEASURES

1. Women make up only 15% of engineers in the field [1].
2. Only 22% of engineering graduates are women [1].
3. The current industry efforts to retain women, in general, have been focused on work climate, providing promotional opportunities, leadership growth, and addressing bias.
4. Women may feel more drawn to enter professional areas that already have a critical mass of women in them [2].
5. Research shows diverse organizations deliver better business results [3].

Now, with a shared understanding of the current situation, you begin the discussion by highlighting that it is the decades-old engineering system that has created the problem and it is now the engineering system that needs to fix the problem. You then lead the group through the myriad of areas that influence the retention of female engineers, writing each of them on the whiteboards behind you from the single sheet of paper you brought to the room. You do this because the first step in solving a problem is to understand the entire problem, including how the problem is currently being approached and what is not working well.

THE CURRENT REALITY OF THE EFFORT

In the Society of Women Engineer's (SWE) review of 2020 literature in their State of Women in Engineering 2021 report, Meiksins et al. write

> Previous SWE reviews of the literature on women in engineering have revealed that there is no consensus as to how to explain the low numbers of women in engineering

> and the frustrating reality that progress toward gender equity has slowed. This year's review is no different.
>
> **[4, p.18]**

Above, we used the description of "myriad" of influences because that is the current conceptualization that we have of the problem. As overwhelming as the sheer number of influences may be, knowing that other formally male-dominated fields, like medicine, have successfully attracted and retained women gives us a case study to learn from and ongoing motivation to work toward a solution in the engineering environment. In SWE's 2021 report, the authors close the introduction section with, "Finally, given the persistent lack of diversity in technical fields, we suggest that there are both unexplored and underexplored areas for future research and intervention" [4, p.18]. Although the problem seems overwhelming, we can feel confident that there are still solutions that can be found and, given the success of other fields, our goal is achievable. While the above literature review by SWE certainly makes the situation sound bleak, we can use each of the areas that researchers have focused on to help us unpack the problem and assess for the underlying root causes of our problem.

Growth Potential and Empowering Work Environment

Growth potential and an empowering work environment matter to women in engineering. In a joint production between SWE and People at Work, the 2019 Women in Engineering Talent Pulse Report describes the top priorities of women engineers in the workplace based on survey results collected from 2,971 respondents with an average of 13.6 years of experience [5]. The most important factors in their decision to remain with their current employer were growth potential (24%) and empowering work culture (23%) [5]. Similarly, training and opportunities for development were listed as one of the five most important benefits of selecting a new employer [5]. The available opportunities and ability for growth and development may be factors in the retention of women in engineering.

Gender Pay Gap

The SWE – People at Work Talent Pulse report also showed that about 70% of the women surveyed said that they were satisfied with their salary while roughly 17% said they were dissatisfied [5]. Further, another study found that only a small percentage of women (less than 10%) identified low salary as a reason for leaving engineering [6]. While a pay gap between men and women remains (female engineers earn 90 cents for every dollar that male engineers earn [7]), the gap is actually less than what has been found to exist between male and female physicians [8, 9] and is better than the overall average in the United States [10]. While work remains to close the gap within engineering and it may be contributing to the problem, it seems less likely to be an underlying root cause.

Inspired to Come to Work

The SWE – People at Work Talent Pulse report leads a section with, "But even with loyalty and salary satisfaction on the upswing, inspiration and motivation at work are low" [5, p. 10–11]. Most engineers choose engineering as a career because they are inspired to solve problems – problems that matter. Successful firms are productive because their employees are inspired and encouraged by their work. In looking at the work women are doing and the environment they are working in, the SWE – People at Work survey found that only "8% feel greatly inspired to do more than is required" [5, p. 12]. In addition, only "9.7% nearly always look forward to going to work" [5, p. 13] while "45.8% of respondents say they look forward to going to work 'about half the time,' 'occasionally,' or 'hardly ever'" [5, p. 13]. A lack of feeling valued and fulfilled at work, or low work satisfaction, likely contributes to the retention of women engineers.

The Type of Engineering and the Type of Work

Although women earn 51.7% of Environmental Engineering degrees, they only earn 15.7% of Mechanical Engineering degrees [11]. While it may appear that a significant number of women are environmental engineers, in 2019 there were 16 times more Mechanical Engineering degrees earned annually than Environmental Engineering degrees earned in the United States [11]. Additionally, a Mechanical Engineering degree is the Number 1 engineering degree earned by women as well as men. The draw of women to Environmental Engineering may be a combination of the social contribution of the work as noted previously and the draw of joining a field with a higher number of female peers. Women currently make up 29% of working environmental engineers but only 13% of mechanical engineers [12, 13]. The difference between Environmental Engineering and Mechanical Engineering indicates that the type of work may play a role in both attraction of women to the field of engineering and retention of women in engineering.

Relational Work

After choosing the general field of work, the actual job or work done within an organization has significant variability. In Joyce Fletcher's 1999 book *Disappearing Acts: Gender, Power and Relational Practice at Work,* the author describes the role that many women engineers play in technology firms – being responsible for the relational needs of product development through relational practices [14]. These responsibilities focus on ensuring that the overall necessary work gets done, that plans are created and then executed, and that nothing falls through the cracks. These responsibilities include ensuring that the teams are able to function in a cohesive manner and that interpersonal issues within the team are addressed or prevented from occurring. Although an organization may desperately need or depend on this work, organizations also typically devalue this work in favor of technical work. Further, women do not always actively choose to be assigned these roles as noted in the 2014 book *What Works for Women at Work*, where the authors describe how women are pressured into classically "feminine" roles and are assigned "office housework" [15, p. 68].

Social Contribution and Communal Work

Research has shown that women prefer to work with people in a career that makes a social contribution while men prefer to work with things in careers with high status [16, 17]. This difference has historically been thought to contribute to the lack of women in the field of engineering; a field perceived as very "things-oriented" and not highly associated with altruism [16]. However, research has also found that differences in values may not change commitment to engineering and that individuals may choose occupations that are not in line with their interests [16, 18]. This indicates that the choice of career path is much more complicated than an individual's underlying values and interests. Further, it is also important to note that an individual's values and interests are shaped by underlying gender schema. One approach to this challenge is to focus on changing the perception of engineering and highlight the ways in which it can help improve society [19].

The Choice for Other Work

Although there have been two decades of effort focused on encouraging girls and women to choose engineering as a career, women have many other career options available to them [2]. Many of these career options are not dominated by men and, therefore, may not come with the risk of negative interpersonal interactions, experiences of being ignored, and the challenges of navigating different work styles. In addition, a portion of women interested in the work of engineering may want different things out of a technical career and the current jobs available in engineering may not meet their needs [20].

Career Path

The career path of women within engineering matters. A study conducted by Cardador and Hill (2018) examined how the career path affected attrition using data from 274 engineers, 40% of whom were women [21]. The study used three previously identified career paths – managerial, technical, and hybrid – to establish the influence of career paths on intent to leave. Cardador and Hill wrote,

> For women, being on the technical path was favorable to the being on the managerial path in terms of intent to leave engineering; the hybrid and technical paths were favorable to the managerial path in terms of identification with other engineers and meaningful work.
>
> **[21, p. 105]**

The authors further note that this counters the retention strategy of promoting women to managerial roles in hope of attracting more women and therefore reducing gender disparity in engineering firms. Based on this study, if we are able to facilitate women maintaining technical responsibilities, we may improve our retention rate.

Women Role Models and Mentors

Role models and mentors play an important role in an engineering firm's work climate, supporting career development, and facilitating professional growth [22]. Having high-level women leaders who serve as role models, or individuals whom other women engineers can emulate, is important. However, it is also important to have role models and mentors who fully understand the day-to-day work responsibility – doing similar work and therefore better understanding the challenges that arise. Studies examining why women leave engineering have commented on the importance of having supportive networks and mentors at work [6].

Work–Life Balance and Enabling Part-Time Work

In a study surveying 5,562 women with bachelor's degrees in engineering, the authors found that 27% of respondents had left the engineering field [20]. Fouad et al. wrote that the "majority of women who left the engineering field stated that it was difficult for them to find part-time jobs in the engineering field, and that was the main reason they left the occupation altogether" [20, p. 5]. These data suggest that designing the engineering system to allow part-time work could have a dramatic effect on retaining women in engineering. However, this would require leaders to make the choice to facilitate part-time work.

Innovation and Patents

A study done by Bell et al. found that, at the time the study was published, approximately 80% of 40-year-old inventors were men [23]. Other studies have found a similar gender difference in patent holders. The National Center for Women & Information Technology (NCWIT) 2016 report described one study that found that 87.4% of information technology patents were created by male-only teams compared to 10.5% by mixed-gender teams and 2.1% by female-only teams [24]. The NCWIT report listed multiple potential factors for this, including women being in execution roles rather than creator roles [24]. Notably, there was significant variability between organizations, with some companies having a much higher percentage of female patent holders [24]. Each organization therefore needs to examine its own internal metrics to understand any discrepancies as well as the main drivers for it. Importantly, though, this demonstrates how shifting women into relational roles may decrease their ability to innovate and also highlights the overall importance of technical depth and skills.

Work Climate

For the past two decades, organizations have focused on improving work climate, reducing bias, and increasing professional opportunities in order to attract and retain women. The focus on establishing a sense of belonging has helped to move engineering work culture in a positive direction and has created a foundation that we now need to build upon. However, SWE reported in 2021 that "30% of women who have left engineering cite organizational climate as the reason" [7, p. 2], indicating that there is still more work to be done.

Summary

In the SWE 2020 State of Women in Engineering report's review of the 2019 literature, the authors have a section entitled, "Change the women or change engineering?" [2, p. 33]. As our goal is not to change women, nor is it realistic or feasible, this leaves us with two options: change engineering or live with the current performance.

Getting back to our story above, we now find that the whiteboards are completely filled with contributing factors. The team counts the items listed on the whiteboards and finds almost 20 relatively independent areas of influence. Each of these is important and can impact women in the workplace, but in different ways. We can imagine that although the team may feel somewhat overwhelmed with all of the information, they may also recognize that there are likely fundamental underlying problems that may not be readily apparent and are therefore missing from the list. This is a little bit like analyzing the problems in a process and ending up with a flat Pareto chart,* showing no tall bars. Because we don't know which causes are contributing the most, it's unclear what to work on first. The question then becomes, with a flat Pareto chart, what is the next step?

DIVERSITY MATTERS

Before we move too quickly to our next steps, let's reiterate the financial reason driving this effort: diversity delivers better business results [3]. We know that this occurs from the executive level down to the individual engineer level. As engineers, designers, and problem-solvers, we know that delivering the best solution requires fully understanding the problem and the need. It requires having a team that is able to look at the entire solution space and think critically about all possible options. It requires having a team that is able to set their biases aside and look at the solution space through the eyes of the customer and the pure technical requirement. Let's face it, a team of five white men who are all in the same stage of their career will find it challenging to deliver the same cutting-edge solutions that a diverse set of problem solvers can deliver.

Up until the early 1980s, the typical relationship between Product Development and Manufacturing was referred to as "Throw it over the wall." Understanding the product delivery issues that arose from this philosophy, and the business cost of

* A Pareto chart, named after Vilfredo Pareto, contains vertical bars that when shown from left to right, includes the most significant cause of a problem on the left and then lesser causes moving to the right. The Pareto principle states that about 80% of the consequences come from 20% of the causes, giving us the 80/20 rule.

this approach, led to a significant amount of work done in the area of Design for Manufacturing (DFM). Engineers with manufacturing experience and responsibilities were integrated directly into the product design process to address the problem. At the most basic level, this expanded diversity within the development team leading to significant financial benefits. By the mid-1990s, if a firm had failed to incorporate DFM into their development process they may have failed to exist or, at a minimum, struggle to compete.

CONCLUSION

In our hypothetical story, we would expect the leader to summarize the situation for the team. The leader might start with how we know that organizational climate has been cited as the cause for 30% of women leaving engineering and that our underlying gender schema plays a role in work climate [7]. We know that having capable women role models will improve the work climate. While women are leaving engineering, we know that they are listing training and development as their top reasons for wanting to stay [5]. We know that a significant number of women are not inspired to come to work [5]. We can assume that encouraging women to remain in technical areas can help some women remain in the field given the differences in retention between technical and managerial paths [21]. We know that women in medicine have seen a very different overall outcome [25]. We know that women may gravitate toward career areas that already have a critical mass of women [2]. Finally, we know that efforts that have been made to address these areas have not been successful at increasing retention of women in engineering and we need to find a new approach.

Next Steps for the Team

Typically, the leader might provide some general direction for the next steps while the team takes pictures of the whiteboards and then heads out of the room. However, in this case, the leader has a very specific request. The leader asks the team to start with the fundamental problem of a lack of women in engineering and then create a causal diagram to break the problem down. In addition, the leader points out that given that half of all students in medical school are women, there is probably something to be learned by viewing medical training as a case study. At the end of our story, for now, after pictures of all of the whiteboards are taken, we can also expect that sadly no one stuck around to clean them off, but maybe that was part of the leader's plan.

REFERENCES

1. R. Fry, B. Kennedy, and C. Funk, "STEM Jobs See Uneven Progress in Increasing Gender, Racial and Ethnic Diversity," Pew Research Center, Apr. 2021. [Online]. Available: https://www.pewresearch.org/science/wp-content/uploads/sites/16/2021/03/PS_2021.04.01_diversity-in-STEM_REPORT.pdf.
2. P. Meiksins, P. Layne, K. Beddoes, and J. Deters, "Women in Engineering: A Review of the 2019 Literature," *SWE Magazine*, vol. 66, no. 2, pp. 4–41, Apr. 2020.

3. V. Hunt, L. Yee, S. Prince, and S. Dixon-Fyle, "Delivering Through Diversity," McKinsey & Company, Report, Jan. 2018. Accessed: Jan. 15, 2022. [Online]. Available: https://www.mckinsey.com/business-functions/people-and-organizational-performance/our-insights/delivering-through-diversity.
4. P. Meiksins, P. Layne, and U. Nguyen, "Women in Engineering: A Review of the 2020 Literature," *Magazine of the Society of Women Engineers*, vol. 68, no. 1, Winter 2022. [Online]. Available: https://magazine.swe.org/women-in-engineering-a-review-of-the-2020-literature/.
5. R. Rincon, and D. Linstroth, "Women in Engineering Talent Pulse Report 2019," People at Work and the Society of Women Engineers, 2019. Accessed: Dec. 03, 2021. [Online]. Available: https://swe.org/wp-content/uploads/2019/10/Compressed-People-At-Work_SWE_Talent-Pulse-Report.pdf.
6. N. A. Fouad, and R. Singh, "Stemming the Tide: Why Women Leave Engineering," Center for the Study of the Workplace at University of Wisconsin - Milwaukee, 2011. [Online]. Available: https://www.energy.gov/sites/prod/files/NSF_Stemming%20the%20Tide%20Why%20Women%20Leave%20Engineering.pdf.
7. Society of Women Engineers, "SWE Research Fast Facts," Society of Women Engineers (SWE), Sep. 2021. [Online]. Available: https://swe.org/wp-content/uploads/2021/10/SWE-Fast-Facts_Oct-2021.pdf.
8. G. Redford, "New Report Finds Wide Pay Disparities for Physicians by Gender, Race, and Ethnicity," *AAMCNews*, Oct. 12, 2021. Accessed: Dec. 24, 2021. [Online]. Available: https://www.aamc.org/news-insights/new-report-finds-wide-pay-disparities-physicians-gender-race-and-ethnicity.
9. A. T. Lo Sasso, D. Armstrong, G. Forte, and S. E. Gerber, "Differences in Starting Pay for Male and Female Physicians Persist; Explanations for the Gender Gap Remain Elusive," *Health Affairs*, vol. 39, no. 2, pp. 256–263, Feb. 2020, doi: 10.1377/hlthaff.2019.00664.
10. A. Barroso, and A. Brown, "Gender Pay Gap in U.S. Held Steady in 2020," Pew Research Center, Washington, DC, May 2021. Accessed: Dec. 24, 2021. [Online]. Available: https://www.pewresearch.org/fact-tank/2021/05/25/gender-pay-gap-facts.
11. American Society for Engineering Education, "Engineering and Engineering Technology by the Numbers 2019," American Society for Engineering Education, Washington, DC, Report, 2020. Accessed: Dec. 03, 2021. [Online]. Available: https://ira.asee.org/wp-content/uploads/2021/06/Engineering-by-the-Numbers-2019-JUNE-2021.pdf.
12. "Mechanical Engineer Statistics and Facts in the US," *Zippia Careers*, Dec. 14, 2021. https://www.zippia.com/mechanical-engineer-jobs/demographics/#gender-statistics (accessed Feb. 01, 2022).
13. "Environmental Engineer Statistics and Facts in the US," *Zippia Careers*, Dec. 14, 2021. https://www.zippia.com/environmental-engineer-jobs/demographics/ (accessed Feb. 01, 2022).
14. J. K. Fletcher, *Disappearing Acts: Gender, Power, and Relational Practice at Work*. Cambridge, MA: MIT Press, 1999.
15. J. C. Williams, and R. Dempsey, *What Works for Women at Work: Four Patterns Working Women Need to Know*. New York: New York University Press, 2014.
16. J. M. Lakin, V. A. Davis, and E. W. Davis, "Predicting Intent to Persist from Career Values and Alignment for Women and Underrepresented Minority Students," *International Journal of Engineering Education*, vol. 35, no. 1(A), pp. 168–181, 2019.
17. R. Su, J. Rounds, and P. I. Armstrong, "Men and Things, Women and People: A Meta-Analysis of Sex Differences in Interests," *Psychological Bulletin*, vol. 135, no. 6, pp. 859–884, Nov. 2009, doi: 10.1037/a0017364.

18. B. Ertl, and F. G. Hartmann, "The Interest Profiles and Interest Congruence of Male and Female Students in STEM and Non-STEM Fields," *Frontiers in Psychology*, vol. 10, pp. 897, Apr. 2019, doi: 10.3389/fpsyg.2019.00897.
19. J. M. Lakin, D. Marghitu, V. Davis, and E. Davis, "Introducing Engineering as an Altruistic STEM Career," *The Science Teacher*, vol. 88, no. 4, Apr. 2021, Accessed: Dec. 06, 2021. [Online]. Available: https://www.nsta.org/science-teacher/science-teacher-marchapril-2021/introducing-engineering-altruistic-stem-career.
20. N. A. Fouad, W.-H. Chang, M. Wan, and R. Singh, "Women's Reasons for Leaving the Engineering Field," *Frontiers in Psychology*, vol. 8, pp. 875, Jun. 2017, doi: 10.3389/fpsyg.2017.00875.
21. M. T. Cardador, and P. L. Hill, "Career Paths in Engineering Firms: Gendered Patterns and Implications," *Journal of Career Assessment*, vol. 26, no. 1, pp. 95–110, Feb. 2018, doi: 10.1177/1069072716679987.
22. S. Dinolfo, and J. S. Nugent, "Making Mentoring Work," Catalyst, New York, NY, 2010. Accessed: Feb. 05, 2022. [Online]. Available: https://www.catalyst.org/wp-content/uploads/2019/01/Making_Mentoring_Work.pdf.
23. A. Bell, R. Chetty, X. Jaravel, N. Petkova, and J. Van Reenen, "Who Becomes an Inventor in America? The Importance of Exposure to Innovation," *The Quarterly Journal of Economics*, vol. 134, no. 2, pp. 647–713, May 2019, doi: 10.1093/qje/qjy028.
24. C. Ashcraft, B. McLain, and E. Eger, "Women in Tech: The Facts," National Center for Women & Information Technology, 2016. Accessed: Dec. 03, 2021. [Online]. Available: https://wpassets.ncwit.org/wp-content/uploads/2021/05/13193304/ncwit_women-in-it_2016-full-report_final-web06012016.pdf.
25. Association of American Medical Colleges (AAMC), "2019 Fall Applicant, Matriculant, and Enrollment Data Tables," Association of American Medical Colleges (AAMC), Dec. 2019. Accessed: Jul. 05, 2021. [Online]. Available: https://www.aamc.org/media/38821/download.

6 Finding Root Cause through a Causal Diagram*

As we saw in Chapters 4 and 5, the underlying causes of the low percentage of women in engineering are complex. In Chapter 5, we identified almost 20 relatively independent factors affecting whether women choose to stay in or leave the field of engineering. In choosing the most important factors, there is a risk of making assumptions based on our own experiences and biases. This could easily turn into an approach akin to handwaving or, worse, handwaving with bias. As individuals and teams, it can be easier to focus on things that seem to be the most within our control, like conducting training about implicit bias.

Sometimes, as problem solvers, we want to find quick fixes that make us feel good for putting in effort and seem like they move us toward a better situation. This works – until it doesn't. When things don't work, we naturally question our dedication to the effort and the amount of energy we are putting in. We either double down on our efforts and try to work harder or quit. It is difficult to take a step back and acknowledge that while what we are working on is important, it may not be the root cause or our method to address it may not be effective. We can draw on a simple analogy as we look at our problem. Imagine that a small town is covered in floodwaters. At first glance, we can see that the water height covering individual streets is an issue, as is the number of homes and businesses that are impacted and the number of acres of farmland under water. However, the issue that matters at the highest level is identifying the cause of the flood. If the flood is caused by a broken levy or a swollen river, the plan of action will be different. Similarly, we need to take a step back and look at the root causes of the low retention of women in engineering.

Let's take a look at a real-world example of addressing the root cause and creating change at a system's level:

---

* There are multiple forms of problem-solving tools, but many, such as those coming out of manufacturing environments, largely deal with "linear problems." Linear problems have generally independent causes and do well with tools like fishbone diagrams (sometimes referred to as an Ishikawa diagram). A fishbone diagram typically does not lend itself well to the problems that have multiple interactions such as our issue with retaining women in engineering. For example, please see American Society for Quality (ASQ), "Fishbone," *American Society for Quality (ASQ)*, 2022. https://asq.org/quality-resources/fishbone (accessed Jan. 17, 2022). In contrast, a causal diagram depicts relationships between variables and can therefore better describe problems with many interrelated factors. An internet search will provide examples.

DOI: 10.1201/9781003205814-9

SOLVING A PROBLEM THAT MATTERED, FINDING THE LONGITUDE [1]

Hundreds of years ago, seafaring captains traveled the world's oceans with the best maps they had at that point in time. Nearly every new exploration improved those maps; however, those early sea travelers had a fundamental problem. At any given time while at sea, they may not know precisely where the ship was located on the map – a map which itself may be inaccurate. For centuries, the navigators knew their latitudinal position (north to south from the equator), but did not have a similar level of exactness for their longitudinal position (east to west from some particular point). This lack of positional knowledge affected their ability to accurately map the positions of land masses and hampered a ship's navigator's ability to guide the ship to their desired destination.

As time progressed, and the work of exploration continued, the maps became better and people began to develop a common understanding of what the world looked like and how best to navigate it. However, stories of ships crashing into land masses that the ship's navigators thought they were avoiding or missing land masses they hoped to find abound in seafaring history. For centuries, brilliant minds developed methods that utilized the celestial bodies to establish the longitudinal position that, while useful on land, was not as useful at sea. Ultimately, it took the development of cost-effective chronometers (clocks) in the late 1700s and the early 1800s to create a fundamentally new way for navigators at sea to best establish where they were and, using a good map, which direction to go. The clock, a tool that could sit on top of a desk, changed sea travel.

With an accurate, seaworthy clock and the knowledge of the local time, using solar noon, a ship's navigator could now reliably establish their precise longitudinal[1] position on an accurate map. To help ships establish the correct local time prior to a sea voyage, major seaports erected time-balls (similar in construction to lighthouses) that could be seen by the ships moored within the port. These time-balls were dropped at exactly 1 PM[2] local time to allow the navigator to reset their ship's clocks.

The combination of good maps, having a tool to know the precise time while knowing the local time, and having a method to calibrate the time on a ship allowed sea travel to become a more predictable process. The clock was the key piece to solving one of the root causes of inaccurate navigation at sea.

1. Establishing longitudinal position draws upon the principle that the spinning rotation of the Earth every 24 hours acts like a clock. If a navigator on board a ship at sea knows the exact time from a fixed point established at the start of a journey, then he can establish his longitudinal position from the exact local time. Eventually, Greenwich, England, was selected as longitude 0 degrees, establishing Greenwich Mean Time as an international standard [1].
2. The time of 1 PM was chosen because the astronomers (on land) might be busy observing the sun at noon as it passed through the local meridian in order to reestablish the local time [1].

FOR OUR PROBLEM OF RETAINING WOMEN ENGINEERS

We may not create something as elegant as a world map, but we may end up with something as basic as a clock and find something as structural, in changing how the work is done, as a time-ball. In the end, we will move from a world of best effort to a better system of engineering.

Longitude, from the Sky or a Clock Sitting on a Desk – a Debate [2]

In our story above, about the invention of a chronometer that was worthy of sea travel, there is another equally significant challenge. Specifically, the decades of debates about whether the solution to finding longitude would come from the celestial bodies or some other method. England's issuing of the Longitude Act of 1714 established prize money, in today's equivalent of millions of dollars, "for a method to determine longitude to the accuracy of half a degree of a great circle" [2, p. 53]. This made the objective "to find the longitude" a European effort with significant financial rewards. The efforts to use astronomy to solve the problem were driven by decades of labor to understand the heavens and the expectation that the solution lay in astronomy [2]. Developing an approach using celestial bodies required first understanding how the bodies moved relative to Earth and then making charts and graphs for a navigator to use. Then, with an understanding of the stars' position relative to Earth came the equally challenging work of establishing methods to take measurements on a ship dealing with swells and weeks of cloud cover. Even with the challenges of an effort based on the stars, individuals as highly respected as Sir Isaac Newton discounted the ability to design a chronometer that met the accuracy requirements at sea. They remained fixed in their belief that any solution suitable for the required accuracy would come from celestial observations [2].

Their concern was not without merit, in order for a chronometer to win the prize, it would have to lose or gain no more than three seconds per day. This would equate to two minutes on a six-week voyage. The most formidable challenges for a clock to deal with included changes in temperature, humidity, and gravity. Ultimately though, the solution did come in the form of a clock from a "self-educated maker of an oversized pocket watch" [2, p. 60] by the name of John Harrison. However, despite developing his first clock to meet the required performance specifications in 1735 (H1), it wasn't until Harrison's H5 clock in 1773 that he earned the majority of the prize – 38 years after his first clock and four decades after Newton's death in 1727 [2].

Do we suggest that our challenge of retaining women in engineering is on the scale of finding the longitude? No. While not knowing the longitude caused thousands of sailors to die at sea, our task is to address the challenges women face in engineering and the business cost of failed designs or missed opportunities. However, what we learn from that effort is that finding the longitude required a change in technical direction and addressing entrenched beliefs, and we should be looking for a similar change in direction for our solution.

A CAUSAL DIAGRAM IS OUR MAP

We can picture each problem solver working to find a solution and navigating the vast number of inputs to our problem by using a causal diagram – no different from every ship's navigator using a map to navigate the ocean with the best map available.

A Causal Diagram

A causal diagram is a graphical system-based approach to breaking down a problem and identifying the underlying root causes. The diagram used to show the interaction of Confidence and Bias, Figure 2.1, is a causal diagram demonstrating a workplace interaction and the interrelatedness of the critical elements involved. Constructing a causal diagram is an iterative process, just like the creation of a world map. In general, to create a causal diagram, we write the problem on the left side of the page and then identify the most proximal and significant contributors – writing those items to the right and relating their relationship to the problem with arrows. We continue to break down each of those significant contributing factors into smaller and smaller components, rearranging them and identifying connections between the components as we continue to work toward the right side of the page and the root causes.

As we see the picture of the problem evolve, we continue to evaluate our process and adjust the causal diagram as we learn. The identified causes may be known through data or may be hypotheses that need to be tested. As we continue through the process of creating the diagram, the objective is to find the underlying root causes, which we assume are limited in number. As we test our hypotheses of the factors contributing to the problem, we expect our diagram will change with new causes added and rearrangements made of the relationships between factors. In essence, we need a map of our problem that is "logically correct" and we need to find a solution, or our "clock," to the underlying root causes of our problem.

Our team from Chapter 5 has now moved to a new room. We see them in front of a whiteboard, starting the process of creating a causal diagram to understand the factors contributing to the Low Retention Rate of women engineers. The team might start by considering three broad categories of factors:

- How people interact
- How the work is done
- What the work is

However, these categories can be viewed through different lenses. Given the importance of job satisfaction in job retention, we will re-frame them through the lens of the three elements of job satisfaction discussed in Chapter 4: work climate, satisfying work, and control over career/career trajectory. While how people interact directly correlates with work climate, how the work is done and what the work is both affect whether the work is satisfying and the overall career trajectory.

IT ALL BEGINS WITH LOW PERCENTAGES

In many ways, the problem of a low percentage of women in engineering is like trying to fill a bucket with a hole in it. For women in engineering, there is a low rate of fill (incoming women engineers) and a high rate of leakage (women engineers leaving the field). Although engineering firms have clear opportunities in the area of retention (slowing down or stopping the leak), they only have limited ability to create more engineers (increasing the rate of the fill). A causal diagram allows us to assess each of the factors contributing to each component of the problem more thoroughly.

"Low Percentage of Women in Engineering" naturally breaks down into two major components:

- *Low Percentage of Women Entering Engineering* (the fill)
- *Low Retention Rate of Women in Engineering* (the leak).

Figure 6.1 shows the initial step of our causal diagram, with each of these components having contributing factors.

For *Low Percentage of Women Entering Engineering*, those logically might include:

- Low Interest in High School
- Low Interest in College
- Changing Major in College

Each of these contributing factors can be broken down even further into other causes including educational climate, identification of students, lack of access to Science, Technology, Engineering, and Math (STEM) experiences, and lack of role models.

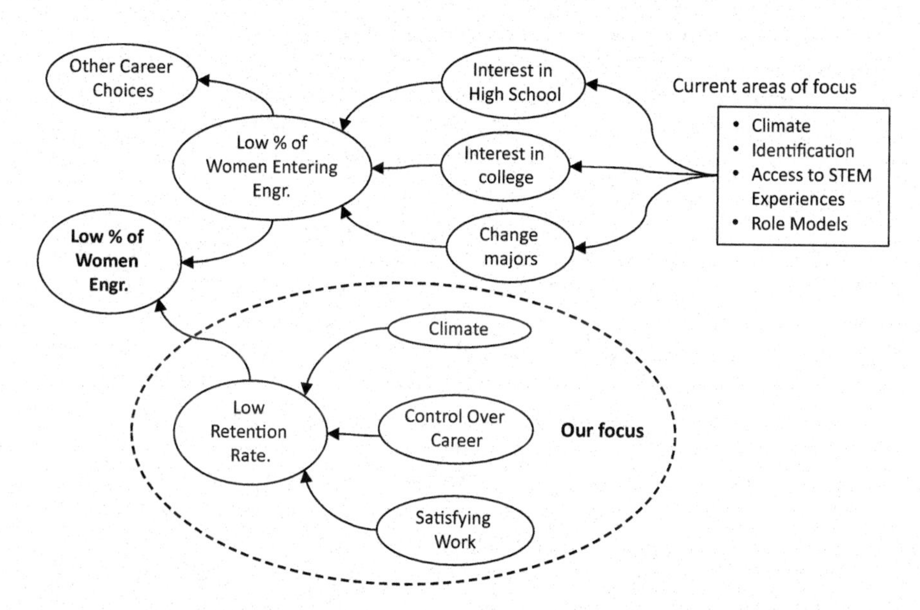

FIGURE 6.1 Partial causal diagram for Low Percentage of Women Engineers.

We also have a "leak" from the overall engineering graduation rate, one in which women graduating with a degree in engineering choose to not enter the field of engineering and make *Other Career Choices.*

While we could continue to go back further into the root causes of each of these factors, given our focus on retention of women in engineering, this is as far as we will take the *Low Percentage of Women Entering Engineering.* We will now turn our focus to the issue of *Low Retention Rate,* as shown by the circled section of Figure 6.1. Our effort, and what a firm can directly control, is to "stop the leak." Based on our summary of the current research in Chapter 5 and our discussion of job satisfaction in Chapter 4, we will begin with three, broad proximal causes of the Low Retention Rate of women in engineering:

- *Poor Work Climate* (this has been supported by research summarized by the Society of Women Engineers [SWE]).
- *Lack of Control over Career* (having Control over Career is a natural expectation of a professional in any career).
- *Unsatisfying Work* (the work may be just a paycheck for some, but work satisfaction may be a critical component to retention for others).

These three areas are fundamental components of any person's career and are gender-neutral. In Chapter 5, we found almost 20 relatively independent areas that affect women in the workplace. From that list, we could have any of the contributing factors, like bias or opportunity, as leading candidates for our proximal causes. However, we soon would have found that we left out a dozen other factors, did not fully identify the relationships between the factors, and, importantly, potentially jumped ahead to a solution all because we became too narrowly focused. The three categories we have chosen as our proximal causes are sufficiently broad.

Based on Figure 6.1, and its limited description listed above, we see a significant level of complexity that is difficult to mentally visualize or fully internalize given all of the interactions between the components. The fact that the human brain takes in vastly more information visually (in the form of pictures) than in written form demonstrates the true value of a causal diagram.

In the development of a causal diagram, it is important to internalize that our objective is to find the most critical components contributing to the problem. Even if it were possible to develop the perfect causal diagram, which is likely not going to happen for a complex problem like this, it is neither necessary nor should be pursued. When utilizing a process tool like a causal diagram, it is critical to understand that there is a point where the work on the process tool needs to stop and the engineering work to solve the problem needs to start. This is true regardless of the process tool used – whether it is a causal diagram or a flow chart. In addition, as we develop our causal diagram for this problem, it is important to internalize that this is presented as one view that, in general, might be 80% correct to 80% of the organizations looking at this problem. It is also possible that those two numbers may be much lower but, in any case, the purpose of the causal diagram we are creating is to identify logical areas for individual firms to consider and to start a dialogue about solving the problem. In addition, because this is about women engineer retention, it may be that an area identified as a primary cause for some women is a non-issue for other women – even within the same firm.

Lastly, as we look at the causal diagram from the perspective of men versus women, we can expect that some factors are specific to women while others may apply to men and women equally. For example, within a firm, *Lack of Control over Career* or *Unsatisfying Work* may be equally important to men and women, but for men the *Work Climate* may be excellent while for women it is not.

Now, to work through, our causal diagram to a detailed level in order to find a root cause.

IT ENDS WITH A LOW RETENTION RATE

Figure 6.2 shows the completed causal diagram* for *Low Retention Rate*. Using the completed causal diagram, we have identified three root causes:

1. *Lack of Reusable Knowledge* (or available knowledge)† (this is supported based on our understanding of training in medicine).
2. *Lack of Role Models* (because this aspect drives both Climate and Control over Career).
3. *Work Methods* (which includes Learning Methods and can reduce Bias and increase Confidence).

We arrived at these root causes by delineating the relationships between the almost 20 issues listed in Chapter 5 and drawing upon our case study of the training environment in medicine. Each of these root causes drives critical elements that contribute to our three overarching areas of *Work Climate*, *Control over Career*, and *Satisfying Work*.

In addition, we have identified five significant influences that need to be addressed as we work through the three root causes. These additional influences are *Bias, Confidence, Technical Coaches, Technical Depth,* and *Balanced Work* (e.g., management of relational work). (Each of the items in the Opportunity Box can be read in the negative sense – e.g., lack of *Role Models*.)

[Although it would be easy to create an additional and simpler view of Figure 6.2, one in which we only show the three root causes and five significant influences, in that view the other critical factors and the impact on other elements would be lost. For example, we are not trying to create Technical Depth just for the reason of doing it, we are expecting that Technical Depth increases Business Contribution and contributes to Personal Growth. Similarly, we aren't creating Reusable Knowledge just to have it, Reusable Knowledge drives Control over Learning which increases the learning rate.

* Figure 6.2 represents a concise view of the causal diagram. A full causal diagram for our issue would include more than a dozen items from Chapter 5. Figure 6.2 was created for both clarity and space considerations.

† For the first part of this book, we wrote of physicians having access to vast amounts of "Available Knowledge" through web-based applications; however, with our eventual introduction of Lean Development, we will change the verbiage to "Reusable Knowledge" with those words being a critical principle of Lean Development.

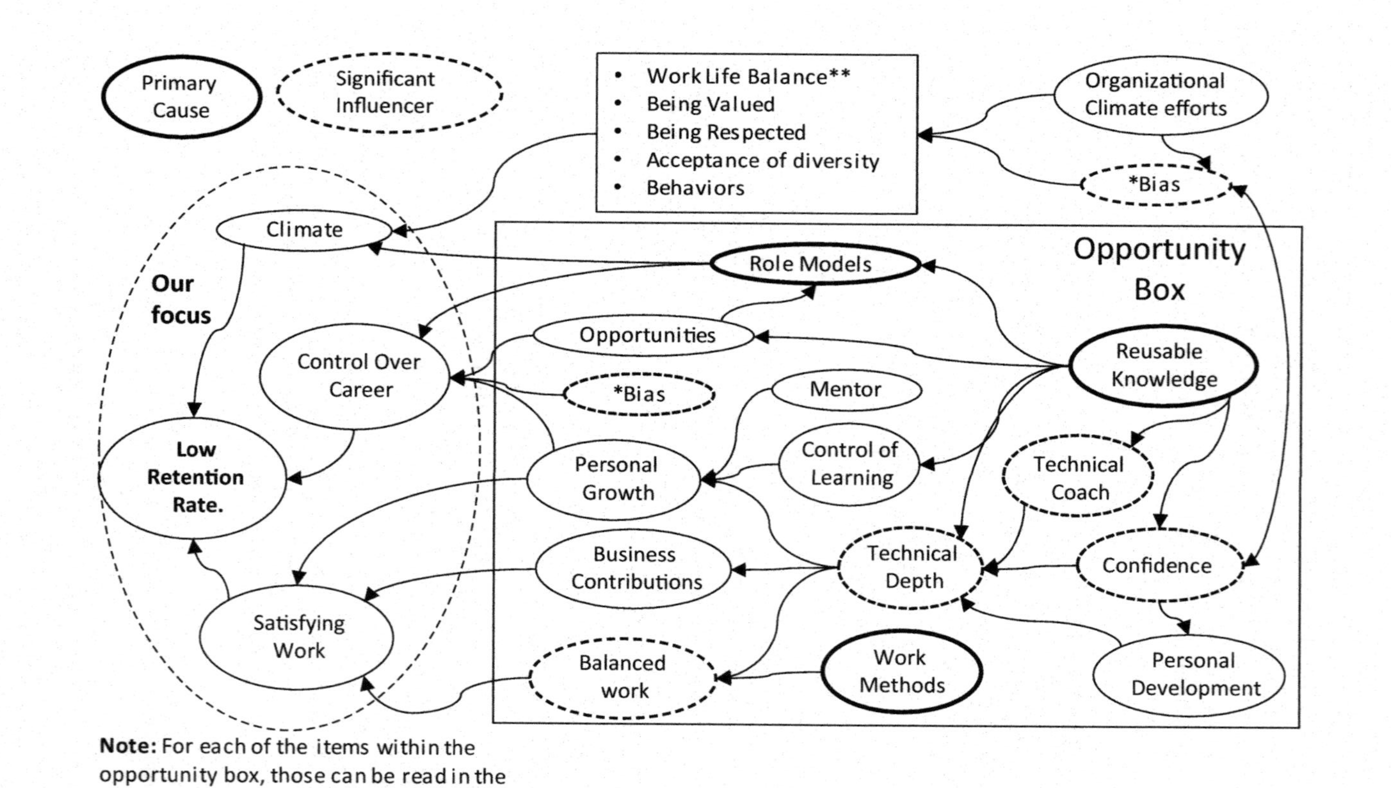

FIGURE 6.2 Causal diagram for Low Retention Rate (*Bias can occur everywhere, ** We include ability for part-time work in this area.).

Therefore, we encourage the reader to spend the time reviewing Figure 6.2 and relate it to their own experiences and perspectives. This figure is also at the end of the book.]

The most critical aspect of the causal diagram and the identified root causes is whether the three causes we have identified are in fact the most basic causes. For the rest of the items, they just need to be logically correct. The Opportunity Box in Figure 6.2 is intended to show all of the major areas that we expect to improve or address with our work on the three primary causes. To assess whether these three causes are indeed the most basic causes, we can ask ourselves the following questions: If we dramatically increase the availability of reusable knowledge, does that contribute to the creation of more role models or increase someone's technical depth? If we change the work methods targeted at doing the work faster and better, does that make the work more satisfying? If we increase the number of role models available, does that improve an individual's control over their career and the work climate? We will address these questions as we go through a detailed review of the key elements of the causal diagram. Before we start breaking down the key factors related to our three larger "buckets" of *Work Climate*, *Control over Career*, and *Satisfying Work*, the following are a few notes about the choices we made in regard to the causal diagram.

As a specific choice, we elected to have *Role Models* as a primary influence over that of *Mentors* because of the uniquely different effects they have on an organization. Although it is reasonable for a person to be both a role model and a mentor, role models within an organization have the ability to motivate a large number of individuals. In contrast, a mentor may be only visible to a few people. In a practical sense, an individual can hire a mentor or coach but would likely not be able to hire a role model.

Additionally, we list the need for *Technical Coaches* to promote *Technical Depth*, because although everyone could benefit from a technical coach, unlike men in which technical coaching may generally occur organically within a firm, this is less likely to occur with women as the minority. In this case, technical coaching needs to be designed into the system. We will cover this area in Chapter 12 and again in Chapters 13 and 14.

Now, we will turn our attention to the components of the causal diagram.

Work Climate

There are multiple ways to break down the climate of a work environment. We will utilize the following six relatively independent areas that can be influenced by a firm.

- *Enabling Work–Life Balance.* This would include the facilitation of someone working part-time.
- *Being Valued.* This includes your paycheck, the role you are in, the type of work you like to do, and emotional factors.
- *Acceptance of Diversity.* At work are you able to be the person you actually are? Does the organization value diversity?
- *Being Respected.* This includes acknowledging the knowledge or skills someone has and demonstrating trust in their abilities. This is beyond just being nice or saying nice things.

- *Behaviors.* This encompasses what is considered professional, acceptable, and expected behavior. This can vary between groups.
- *Availability of Women Role Models.* Positive role models can be a numbers game but are also influenced by the skill set, relatability, frequency of contact, and leadership capability of those who might be role models.

Each of these factors is intertwined with the others, some directly and others indirectly. In addition, some could naturally be combined, for example, being respected and accepting diversity share common attributes. All are important pieces in establishing an inclusive work environment that enhances the success of women and all are influenced by the fact that women are a minority in engineering. For example, the 10% minority (women) are never going to have access to the same number of same-gender role models that the 90% majority (men) have. We have put the first five factors in a separate box because, typically, they are already areas of focus within a "progressive firm." This allows us to simplify the causal diagram so that we can find root causes that are outside of these areas of focus.

Lack of Control over Career

It is a common expectation in most professions that although there are many people who will help an individual achieve their career goals, the individual ultimately "owns their career." They are responsible for understanding what they want out of their career (e.g., just a paycheck, being happy, making a valuable contribution to something that matters, or making it all the way to becoming CEO). In addition, the individual is responsible for establishing the path to that goal and then working toward it. However, it is also the responsibility of the firm's management team to create a work environment that enables someone to own their career and allows them to achieve their full potential. This is what we will be working toward.

Just like with a work climate, there are many ways to break down the management of a career. We will break it down into the following four areas:

- *Personal Growth*, which is driven by *Control of Learning.* This can be simply the career path you desire and how you would achieve it.
- *Opportunities*, which are driven by both the growth of an organization and support for women to pursue them. Having access to reusable knowledge allows someone to recognize available opportunities and take advantage of them, *Reusable Knowledge* therefore increases *Opportunities.*
- *The Impact of Bias.* Bias can affect every aspect of the work. This can range from who is allowed to create a positive work–life balance, to the assigned work (Figure 2.1), to personal development, and to who is promoted as a role model. For the purpose of the causal diagram, assume that Bias is everywhere but we will only show it connected to *Control over Career,* which may be the most overtly damaging place for it to exist for women (e.g., the glass ceiling).
- *Role Models.* In addition to role models improving the work climate, they also demonstrate what is possible for others in the current work environment.

Unsatisfying Work

Relative to retention, how we view our work is largely driven by how satisfying it is. The ability to create satisfying work is firmly within the control of the firm. At its most basic level, we need to differentiate between work that:

- Adds value to a customer
- Work that adds no value to a customer*

Although there are huge preference differences between individuals in regard to choosing work that adds customer value, virtually no one wants to do customer "non-value-add" work. Think of it this way, your manager walks up to you and says, "I have a job assignment for you that adds no customer value but for some reason the organization seems to think it needs to be done, when do you think you can start?" The non-value-add work is generally the type of work that seems to be in place because no one has taken the time to eliminate it. As an individual, it is a pretty simple test at the end of the day: what value did I add for my customer, did I like the work I was doing, and what did I learn? An additional key element of satisfying work is "Does my organization value it?"

When a firm is evaluating engineer retention as a whole and the influence of work, it is important to ensure that there is a clear understanding of the work as it relates to value-add and non-value-add. In addition, for an early-to-career individual, it is critical to ensure that the work has a good balance to enable personal growth and doesn't pigeonhole them, woman or man. In the area of technology, that might be achieved by an appropriate balance between detail design and system-level design or an appropriate balance between customer-centric responsibilities and technical design responsibilities.

To break down *Unsatisfying Work,* we have identified the following:

- *Lack of Personal Growth.* This affects both the work you do and your career path. *Personal Growth* is shaped by *Control of Learning, Technical Depth*, and *Mentors*
- *Low Business Contributions.* This speaks specifically to value-add versus non-value-add work, specifically, when someone is asked to work in an area that does not have a clear business or customer benefit. An individual's *Technical Depth* impacts the business contributions they are able to make.
- *Unbalanced Work.* This is an example of what the work is. For example, an individual who is focused almost exclusively on one type of work is therefore failing to broaden their work knowledge. In a physician's residency program, this is addressed through defined rotations working in multiple areas of work

* From a process management standpoint, we would add a third category of work, Non-Value Add but Necessary. This is the non-value-add work that is necessary to enable the desired performance of the process in its current state of operation. An example of this in a manufacturing process is an End of Production Line test that only validates that the product performs to defined specifications. Without that test, the customer would receive some portion of defective product. See Appendix A for more detailed information.

(e.g., for a specialty of Internal Medicine, it is spending time working in both an intensive care unit and an outpatient clinic). *Poor Work Methods* and lack of *Technical Depth* both contribute to *Unbalanced Work. Work Methods* describe how the work gets done. In Chapter 10, we will introduce how Lean Development approaches the work of product development differently from traditional development and can change how the work is done.

Going a Step Further

The factors listed above for each of our three larger "buckets" of *Unsatisfying Work, Lack of Control over Career,* and *Work Climate* have their own contributing factors which are delineated on the causal diagram. At this point in our process, each of these is a hypothesis about the level of impact. Although each is present within an individual's work and in most cases is driven by the work environment or organizational processes, the amount of impact they have on how someone feels about their job and whether they want to remain in their current job/field varies.

For example, *Technical Depth* contributes to personal growth, business contributions, and balanced work. In our use, the term *Technical Depth* refers not only to pure engineering technical depth but also to how it drives detailed business depth and how it drives system-level technical choices. The impact that technical depth has on an individual's feeling of personal growth may vary.

In addition to *Technical Depth*, other contributing factors include *Lack of Reusable Knowledge (or Available Knowledge)* which comes from our learning from the medical field, *Implicit Bias* and *Confidence* as demonstrated by Figure 2.1, *Personal Development*, and *Technical Coaches.* These factors are related to each other as well as to the more proximal factors listed above.

Arriving at Our Root Causes

From the basic framework in Figure 6.2, we can examine how one element influences another. Certainly, we could find more contributing causes and interrelatedness and end up in a never-ending loop of adding in causes and rearranging our diagram. However, the value of a causal diagram is not to map out every permutation of influence because all of the 15 items in our Opportunity Box are interrelated at some level. The value of the causal diagram, in our case, is to identify the "structural elements" that are most significantly causing the "problem."

The second element of the causal diagram is to check to see if this provides a holistic picture of the issue. Are there any glaring omissions that drive the work or the work environment? Suppose we return to our understanding of Gladwell's three items of satisfying work from Chapter 4, Autonomy, Complexity, and Connection between Effort and Reward, and relate it to our causal diagram in Figure 6.2. In that case, we can assume that *Reusable Knowledge* and *Technical Depth* support Autonomy and Complexity. Additionally, we can assume that *Business Contribution* supports Connection between Effort and Reward. Therefore, we can be reassured that we have a reasonably holistic picture of the issue and are not missing any major factors that drive the work and work environment.

At this point, we can be comfortably certain that our causal diagram is sufficiently mapped out and that we have arrived at three root causes:

- Lack of Reusable Knowledge
- Lack of Role Models
- Work Methods

This is therefore where the process tool stops and the engineering starts.

THE SOLUTION

From Figure 6.2, we can see that if we can make progress in the area of reusable knowledge, we will improve multiple areas beyond that. We can expect to see increased confidence which reduces Bias, we can create stronger (or more) role models, and we can enable personal development plans that lead to increased technical depth resulting in more satisfying work. In addition, reusable knowledge leads to Control over Learning, Personal Growth, and Control over Career.

In the area of product development (which means creating anything new to solve a problem), reusable knowledge comes in two basic forms.

1. *Knowledge that is created by others* and used by engineers and managers doing the work. This could be knowledge at both the corporate level within a firm (involving hundreds of engineers) as well as knowledge created by peers (maybe involving a team of a dozen engineers). In the case of medicine, the vast majority of knowledge available to physicians might be considered industry-level knowledge. This knowledge is available to every physician.
2. *Knowledge created by the individual who is doing the work* and the creation of it as part of doing the work. This is the very deliberate effort by an engineer or manager to take the time and do the work such that she or he is creating their own knowledge base in a way that can be used by others. It is not additional work. It is work that someone could share with a team, mentor, manager, or technical coach in an effort to help someone actually do the work in a more effective and efficient manner. The creation of it has a net positive outcome. To be clear, it is not well-written notes in a personal engineering note book.

We will focus our efforts throughout the remainder of this book on Number 2, knowledge created by the individual who is doing the work – in part because this is controlled by the individual. In Chapter 10, we will describe a very basic tool (the A3 Problem-Solving Tool) that is an industry standard for creating reusable knowledge while simultaneously doing the work of solving the problem. The A3 Problem-Solving Tool can be used at any level of the organization.

This tool, an A3 Report, which is one side of a single sheet of paper (11 × 17 inches in the United States and A3 size paper in the rest of the world), fundamentally drives our effort to create reusable knowledge. The power of this tool is in enabling clarity and conciseness while guiding resolution to the problem. Through a deliberate focus on creating reusable knowledge by an individual doing the work (Number 2), it will naturally create reusable knowledge available to others (Number 1).

Additionally, we will address *Work Methods* through a discussion of Lean Development with four fundamental areas of focus:

- The internalization of Value-Add versus Non-Value-Add
- The importance of reusable knowledge
- The learning process through the development work
- Working at the system level

Creating *Reusable Knowledge*, through A3 Reports, will naturally enable the growth of *Women Role Models* by allowing them to develop stronger technical knowledge and skills in more areas of development within the firm. With the use of A3 Reports by those doing the work, role models can develop an increased understanding of the work that engineers, both men and women, are doing. The role models will develop deeper and broader capabilities. We will discuss role models in detail in Chapters 12 and 14.

CONCLUSION

Using our understanding of the contributing factors to the low retention of women in engineering as well as the factors contributing to the success of our case study (the field of medicine) that we explored in the first five chapters, we have been able to develop causal diagrams. We began with three distinct considerations for the work environment and what it delivers: how people interact, how the work is done, and what the work is. However, given the importance of job satisfaction in job retention, we reframed these three considerations into three broad categories: Work Climate, Control of Career, and Satisfying Work. Using these three overarching areas as our most proximal causes, we were then in a position to utilize a basic engineering tool – a causal diagram – to identify the root causes of our problem with the ultimate goal of finding solutions that can be created through the structural change of the engineering system. Using this iterative problem-solving tool, we identified three root causes: Role Models, Work Methods, and Reusable Knowledge. We additionally identified five contributing factors that affect multiple areas within the causal diagram: Bias, Technical Depth, Balanced Work, Confidence, and Technical Coach. With the causal diagram, we have now found our map; however, just like finding longitude rested on the development of a seaworthy clock, we are still in need of the clock to solve our problem.

REFERENCES

1. D. Howse, *Greenwich Time and the Longitude.* London: Philip Wilson Publishers Limited, 1997.
2. D. Sobel, *Longitude: The True Story of a Lone Genius who Solved the Greatest Scientific Problem of his Time.* New York, NY: Penguin Books, 1996.

Part II

Summary

Using our understanding of the situation described in Part I, we began the process of analyzing the problem in Part II.

- In Chapter 4, given the increase of women in medicine, we examined the basic differences and similarities between the two professions of medicine and engineering. This review led us to one clear difference: access to available knowledge. We explored factors that affect an individual's basic expectation of job satisfaction: work climate, satisfying work, and control over career.
- In Chapter 5, we identified the myriad of areas that either contribute to women staying in engineering or are the reasons for them leaving. The nearly 20 areas we identified have a wide variation in their influence on individual women – mattering for some women and not for others.
- In Chapter 6, in our causal diagram (Figure 6.2), we integrated three areas: what the data says about why women leave or stay in engineering, the work environment of a physician in residency, and our desire to make structural changes to the work and its environment. We identified lack of satisfying work, lack of control over career, and a negative work climate as the three broadest areas of contributing factors. Using the causal diagram, we then identified three root causes: lack of reusable knowledge, work methods, and lack of role models. Each of these is a structural element of the work environment.

DOI: 10.1201/9781003205814-10

Through the remainder of this book, we will focus on changes within the engineering system that target our three root cause areas and the other areas of influence.

As leaders read through the remainder of this book, it may be helpful to outline actions within each of these areas that are specific to your organization with the ultimate goal of creating a causal diagram specific to your organization but that clearly includes the use of reusable knowledge.

Part III

Developing a Solution

Lean as a Foundation for Change and Learning

7 Industry Efficiency via Lean*

Using our causal diagram, Figure 6.2, we have identified three fundamental causes impacting the retention of women engineers:

1. Lack of reusable knowledge
2. Lack of role models
3. Work methods

Through our work with the causal diagram, we know that these three areas can drive or influence bias. Further, we have discussed how current efforts to address bias by focusing on how people interact have not been entirely effective. It is therefore time for a new, systems-based approach that changes the system of engineering. One approach to changing the system of engineering is to change how the work is done, specifically through the implementation of lean processes. By changing how the work is done, behavior change can follow. In order to change the way work is done to effectively target our root causes, we must first understand how we ended up here, how work is currently being done, and in what direction it is currently headed.

A BRIEF HISTORY OF LEAN – THE ELIMINATION OF WASTE

In 1950, Japan had near-zero percent of the world motor-vehicle production while North America commanded 85% of it. Twenty years later (1970), the North American share had dropped to 33%, while Japan provided 14% of the world motor-vehicle production and met 4.2% of North American demand for motor vehicles through exports to the United States [3]. However, even with that shift, little focus was placed on what would soon be a formidable competitor. Over the course of the next decade, the North American automotive companies saw an increase in the real world of competition in the North American market, which they could no longer control or ignore. During that time, the North American autocompanies had to deal with rising gasoline prices due to an oil crisis, which resulted in increased demand for smaller cars which North American car companies were not prepared to manufacture. By 1980, Japan had 24.6% of the world's production while North America dropped further to 25.2%. Of much more concern though, Japan now filled 22.8% of the North American automotive demand [3]. For the first time in US automotive history, US car buyers had

* Lean can simply be characterized as the elimination of waste. The term lean production was coined by John Krafcik in a Fall 1988 Sloan Management Review Article [1]. The book, *The Machine that Changed the World* (1990) [2], provided broad visibility to the use of lean within production, across the enterprise, and within design.

DOI: 10.1201/9781003205814-12

a significant breadth of car choices beyond what was internally produced [4]. The challenges faced by the US automotive industry spurred research efforts to understand how Japanese automotive companies, specifically Toyota, had established such a strong market presence in such a relatively short time. Each piece of research conducted showed that virtually every aspect of Toyota's manufacturing system was done in a different and more powerful way than traditional US methods.

Visibility to these methods came from books such as *A Study of the Toyota Production System* (TPS) (1981) in which the author articulated the purpose of the TPS as "a system for the absolute elimination of waste" [5, p. 67]. The book *Japanese Manufacturing Techniques* (1982) provided visibility to the manufacturing flow of just-in-time production with total quality control [6]. Edwards Deming's book *Out of the Crisis* (1982) established his 14 points as the basis for the transformation of the American Industry [7]. The Japanese published book *Managerial Engineering* (1983) articulated engineering methods utilized in Japan to improve productivity and quality [8]. After recognizing that waste exists within every process, the need for a process to eliminate this waste was understood by those outside of Japan. By the mid-1980s, US industries' desire to utilize Japanese manufacturing methods had spread beyond automotive and created a new focus on delivering products. By the mid-1990s, with the publication of *The Machine that Changed the World* (1990) [2] and *Lean Thinking* (1996) [9], the terminology of lean was engrained as the word of choice for a process to eliminate waste.

For individuals and leaders who are part of struggling businesses, the principles and methods of lean create a foundation to work from. They can give people hope. They give leaders a playbook to work from and most importantly they provide a set of organizational rules and values that leaders can lead from. The implementation of these organizational strategies is referred to as a company's lean transformation.

Dozens of books have been written articulating the methods of lean while highlighting a multitude of lean transformations that have occurred over more than three decades of efforts. These examples range from small companies or small organizations within larger processes to highly visible efforts such as Ford's turnaround beginning in 2007 [10] (while Chrysler and General Motors filed for chapter 11 bankruptcy in 2009).

A very customer-centric, or more specifically patient-centric, success comes from the book *Lean Doctors* (2010) in which the authors describe the application of lean principles in a medical facility, the Orthopedic Center at Children's Hospital of Wisconsin. The transformation through lean principles resulted in a reduction of patient wait times by 71%, a reduction of needed exam rooms by 25%, and an increase in patient volume by 25% [11]. In addition, staff satisfaction scores increased by more than 25% [11]. At the center of the effort, the authors write of two key initial strategic decisions: (1) "Begin One Doctor at a Time" [11, p. vii] and (2) "Focus on Patient Wait Times" [11, p. vii].

With respect to complete transformation efforts or even just the significant improvement of one key business measure, it is fair to assume that there are as many failures as successes. However, through the rest of this book, we will articulate what we expect to be a lean transformation with a positive outcome.

LEAN EFFORTS IN ALMOST EVERY INDUSTRY

By the mid-2000s, with highly visible books published such as *The Toyota Way* (2004), lean efforts were being made across almost every industry [12]. However, in order to understand lean and specifically waste, we must first appreciate its opposite – value, or what the customer wants and will pay for. At a consumer level, we might view this as a competitive product that is both fairly priced and meets our need. An example of waste, then, would be the features that a customer wouldn't pay for or doesn't want. Waste is inherent in every process we create or utilize. Lean methods focus on minimizing the cost and maximizing the value delivered to the customer through the elimination of waste. Applying lean methods to a process allows for not just a 10% or 20% improvement in the delivery of value, but a greater than 50% improvement. Once these improvements have started, opportunities for further improvement continue to build.

WASTE IN A NORMAL PROCESS IS A GREATER THAN 50% OPPORTUNITY.

In the application of the methods of lean:

- If you have a process that is running as anyone might expect to see it run
- Without prior efforts to eliminate waste
- Apply the principles of lean to that process
- You should expect nothing less than a 50% improvement in efficiency or overall value to the customer or business
- Then, from the first 50% improvement, start working on the next 50%

The experience of seeing a greater than 50% improvement through lean is demonstrated in a 2021 article written by Art Byrne, author of *The Lean Turnaround* (2013), in which he writes that after his company, Wiremold, acquired a company,

> on the first day, we would give the new management teams the following list of operational excellence targets to focus on: 100% on-time customer service, 50% reduction in defects annually, 20% productivity gain annually, 20x inventory turns, Visual control and 5S everywhere.
>
> [13] **(5S is a lean method focused on organizing the work area.)**

Each of these is measured by the people doing the work. These targets are in contrast to typical business goals such as increasing sales or ROI (Return on Investment). Similarly, in the book *Scrum, The Art of Doing Twice the Work in Half the Time*, which is grounded in lean and typically used in software development, the authors write, "It's hard to believe, but we regularly see somewhere between a 300 to 4000% improvement in productivity among groups that implement Scrum well" [14, p. 34]. In addition, we also see this greater than 50% improvement in our example from *Lean Doctors* above [11].

Lean production has been the attention of significant research and publications with one of the first and most visible being the publication of *The Machine that Changed the World* in 1990 [2]. However, the roots of lean go back over 100 years to when Henry Ford began to integrate his methods of production with the first moving assembly line, which was instituted at Ford's Highland Park Plant in 1913 [15]. In 1926, Henry Ford published *Today and Tomorrow* and communicated the ways in which his dominance of the US automotive industry was the result of his company doing the work differently [15]. In his book, Ford described specific methods to eliminate waste in the manufacturing system of his Model T. He wrote of significant efforts to save timber, which was largely used to make shipping containers for automotive parts and of the technical challenges of moving from batch production of plate glass to continuous flow, which resulted in significant reductions in factory space, increased capacity, and reduction in manual handling activities [15]. In each of his many examples, he wrote of how the elimination of waste within the production process using mass production resulted in improved business performance and direct benefit to customers through the reduced cost of his automobile. Over the course of Ford's 18-year production of Model T, beginning in 1908, the price of the car was reduced from $850 to less than $300 in 1925 [16].

Similarly, Taiichi Ohno, in his book *The Toyota Production System: Beyond Large-Scale Production* (published in 1978 in Japanese with an English version published in 1988), describes Toyota's methods of success in the US automotive market [17]. Ohno describes the seven areas of waste and specific methods to eliminate waste, and then takes the reader of his book through a methodical description of the Toyota Production System (TPS). He covers every aspect of what is needed to create a world-class manufacturing process with the focus of the effort being first on the identification of waste and then on the elimination of it [17]. Toyota built upon the initial methods of production flow created by Ford, but shifted from the use of mass production to achieve lower costs to the use of small lot sizes. By focusing on small lot size, quick set-ups, and utilizing the method of just-in-time, they reduced waste in areas such as overproduction. Using a systems-level perspective, through the efforts of lean, creates an opportunity for a business.

LEAN DEVELOPMENT*: EFFICIENCY OF CREATING KNOWLEDGE

Given the benefits of a systems-based implementation of lean methods demonstrated in a number of industries, we will now apply a similar systems-based approach to change how the work in engineering is done in order to address the root causes of low retention of female engineers, specifically addressing how engineers learn and their limited access to reusable knowledge as well as creating a positive work

* We use the term Lean Development in the very general sense of creating virtually anything new, including a product, process, or service that is developed for a customer to meet a customer need. The development work could be a product that ranges from a physical thing that a customer holds in their hand to a written report that provides detailed information on market conditions and recommendations to accelerate market growth to a process to improve the efficiency of a production system. The industry may also use the terms lean product development or lean product and process development.

climate. To do this, we will look to Lean Development, whose sole purpose is to efficiently create and deliver a new product or service to someone who is willing to pay for it. A new product or service is created through the creation of knowledge, which might be very technical details needed for the design of a product or service or it could be understanding what customers really want or need or don't even know they need (e.g., a smartphone) but are willing to pay for. A firm's success in the area of product development is therefore based on its rate of creating knowledge that is relevant to the value-add for a customer. The business by-products of that faster rate of learning are lower costs, better designs, higher customer value, increased innovation, and reduced time to market. There are dozens of authors who have focused for the last two decades on the basic question: Why does Toyota continue to dominate the automotive sector? The answer, at its most basic level, is that they create value-add knowledge at a faster rate than their competitors and they are then able to turn that knowledge into profitable products (revenue streams).

Authors note: Bob Emiliani writes extensively about the need to understand why the lean transformation attempts of many firms fail. Our belief is that the two strategies identified in Parts IV and V provide a solid foundation to enable a firms successful transformation.

CONCLUSION

Given the need for a systems-level change to address the root causes of low retention of women in engineering and one of those root causes being a lack of reusable knowledge, we have turned our attention to a process that will change the way the work is done. Lean processes focus on the elimination of waste and increasing value add. Implementation of lean processes has been associated with significant improvement, greater than 50%, and has been applied in many different areas of industry. Lean Development, specifically, is a process that focuses on increasing the rate at which knowledge is developed in order to deliver a better product faster and could therefore be used to target all three of our root causes – increasing reusable knowledge, changing the work methods, and increasing the number of role models. Before we discuss how to apply Lean Development processes to the field of engineering, we will first turn to our case study to develop a better understanding of the learning process in the field of medicine.

REFERENCES

1. J. F. Krafcik, "Triumph of the Lean Production System," *MIT Sloan Management Review*, vol. 30, no. 1, pp. 41–52, 1988.
2. J. P. Womack, D. T. Jones, and D. Roos, *The Machine that Changed the World: How Japan's Secret Weapon in the Global Auto Wars will Revolutionize Western Industry*. New York: Rawson Associates, 1990.
3. A. A. Altshuler, M. Anderson, D. Jones, D. Roos, and J. Womack, *The Future of the Automobile: The Report of MIT's International Automobile Program*. Cambridge, MA: MIT Press, 1984.

4. From 1979 to 1981, Bob was able to see this competition first-hand while working as a lot boy (e.g. moving and washing cars) for new car dealership which sold Oldsmobile (General Motors) and Toyota vehicles.
5. S. Shingō, *A Study of the Toyota Production System: From an Industrial Engineering Viewpoint*, Rev. ed. Cambridge, MA: Productivity Press, 1989.
6. R. Schonberger, *Japanese Manufacturing Techniques: Nine Hidden Lessons in Simplicity*. New York: Free Press, 1982.
7. W. E. Deming, *Out of the Crisis*. Cambridge, MA: Massachusetts Institute of Technology, Center for Advanced Engineering Study, 1982.
8. R. Fukuda, *Managerial Engineering: Techniques for Improving Quality and Productivity in the Workplace, 1*. Engl. ed. Cambridge, MA: Productivity Press, 1983.
9. J. P. Womack, and D. T. Jones, *Lean Thinking: Banish Waste and Create Wealth in Your Corporation*. New York: Simon & Schuster, 1996.
10. J. M. Morgan, and J. K. Liker, *Designing the Future: How Ford, Toyota, and Other World-Class Organizations Use Lean Product Development to Drive Innovation and Transform Their Business*. New York: McGraw-Hill, 2019.
11. A. Suneja, and C. Suneja, *Lean Doctors: A Bold and Practical Guide to Using Lean Principles to Transform Healthcare Systems One Doctor at a Time*. Milwaukee, WI: ASQ Quality Press, 2010.
12. J. K. Liker, *The Toyota Way: 14 Management Principles from the World's Greatest Manufacturer*. New York: McGraw-Hill, 2004.
13. A. Byrne, "Ask Art: What Targets Should We Set When Launching a Lean Turnaround," *Lean Enterprise Institute*, Aug. 25, 2021. https://www.lean.org/the-lean-post/articles/ask-art-what-targets-should-we-set-when-launching-a-lean-turnaround/ (accessed Nov. 30, 2021).
14. J. V. Sutherland, and J. J. Sutherland, *Scrum: The Art of Doing Twice the Work in Half the Time*, First Edition. New York: Crown Business, 2014.
15. H. Ford, *Today and Tomorrow: Commemorative Addition of Ford's 1926 Classic*. Boca Raton, FL: CRC Press, 2003.
16. Britannica, The Editors of Encyclopaedia, "Model T," *Encyclopedia Britannica*. May 29, 2020. Accessed: Nov. 11, 2021. [Online]. Available: https://www.britannica.com/technology/Model-T.
17. T. Ohno, *Toyota Production System: Beyond Large-Scale Production*. Boca Raton, FL: CRC Press, 1988.

8 Creating a Physician

The Learning Process

To facilitate our use of training in medicine as a case study for training in engineering, this chapter will provide an overview of a physician's training. The training process includes obtaining a bachelor's degree, completing prerequisite courses, graduating from medical school (typically after four years of study), and completing specialty training through residency and a possible fellowship. There are regulatory agencies providing input and standards throughout this process including the Liaison Committee on Medical Education (LCME) and Accreditation Council for Graduate Medical Education (ACGME) that oversee and accredit medical schools and graduate medical training programs.

TRANSFORMATION OF PHYSICIAN TRAINING

To fully understand the current model of training physicians, we have to look back at the development of medicine as a field and the development of medical training over time. Beginning in Ancient Greece, in the era of Hippocrates in the 5th-century BCE, students of medicine followed the apprenticeship model [1]. They learned directly from a master with an emphasis on learning medicine at the bedside from patients. Notably, during this time, access to reusable knowledge in the form of texts was limited as texts had to be copied by hand. As the first European medical schools were created in the early 11th-century CE, medical education transitioned from practical training at the bedside to an academic approach, focusing on careful analysis and understanding of texts written by classical and Islamic authors [1]. Meanwhile, disease frameworks and texts on treatments were simultaneously developing in the non-Western world. With the transformation of scientific understanding, including germ theory, and the development of new laboratory techniques, the emphasis shifted again – this time to research and medical science [1].

While medical training and medical sciences had been developing for centuries in Europe and Asia, the first medical school in the United States was created in 1765 at the College of Philadelphia [2]. During the remainder of the 18th century and throughout the 19th century, an increasing number of medical schools opened with variable clinical training as there was no standardized clinical instruction and no guaranteed clinical exposure or facilities [3]. These medical schools, termed proprietary schools, were businesses with some focused primarily on the profit margin [3]. To obtain additional training and clinical experience, future physicians found alternate routes. One option after completing medical school was to participate in an apprenticeship which allowed for clinical experience and individualized learning [3]. However, apprenticeship as a training model was limited by variability in the

DOI: 10.1201/9781003205814-13

quality of the teaching physician, access to new and developing medical knowledge, and breadth of training [3, 4]. For those with the financial means, another option for clinical experience was to attend an extramural school during the summer when the proprietary schools were not in session [3]. Alternatively, for a few individuals, there were a small number of hospitals with formal training systems, which turned the apprenticeship into a formalized internship [3]. However, given the number of new doctors in the United States and limited available spots, medical school graduates continued to look for additional clinical experience and internships, with many traveling to Europe, particularly France and Germany, to gain further knowledge [3, 4].

While post-graduate education required expansion in the United States, the focus was placed first on improving undergraduate medical education in response to concerns raised about public safety in light of practicing physicians with varying degrees of training and skill [3, 4]. Beginning in the mid-19th century and throughout the early 20th century, the medical education system underwent significant reforms including new entrance requirements, changes in teaching methods, and changes in the organizational structure of medical schools. These changes contributed to the creation of the modern medical school [3]. While this reform process was occurring, the number of proprietary schools was declining. Proprietary schools were ultimately eliminated through a combination of state licensing laws introduced in the late 19th century and the publication of the *Flexner Report* in 1910 which offered a critique of the state of medical schools in the United States [3, 4]. By the 1920s, with increased regulation, the closure of many medical schools, standardization of medical education, and the creation of the modern medical school, the current undergraduate medical education system was in place [3].

Meanwhile, in post-graduate education, the first modern residency program was created at Johns Hopkins University, representing a transition from vocational training to graduate education [4]. At Johns Hopkins, there was an emphasis on how to acquire and evaluate information rather than on acquiring a body of knowledge as well as a tradition of increasing levels of responsibility while progressing through residency [4]. The number of residency programs increased throughout the first half of the 1900s, with many program leaders having completed a residency at Johns Hopkins [4]. Concurrently, a shift in the understanding of disease processes led to recognition of diseases as arising from particular organs and tissues, leading to the proliferation of specialties within medicine and a resulting need to regulate who was qualified to practice specialty medicine [1, 4]. This was followed by the rise of specialty boards, the first being the American Board of Ophthalmology in 1917 [4]. The function of these boards included knowledge examinations to test candidates' knowledge and abilities before certification and specification of required prior experience for eligibility to take the exam [4]. Similarly, regulations with requirements for training were established to approve residency programs, including aspects like gradually increasing responsibility in patient care, supervision by qualified instructors, educational components, and adequate facilities [4]. Residency programs are now monitored and accredited by the Accreditation Council for Graduate Medical Education (ACGME) and physicians are required to have completed an accredited residency program, taken the required specialty certification exam, and obtained a full medical license in the state they are practicing in before they can become board-certified specialists [5, 6].

THE ROAD TO BECOMING A PHYSICIAN

While there are many different paths to becoming a board-certified physician, there are key points that each path has in common to meet licensing requirements. The typical journey starts with completing a four-year degree, prerequisite courses, and taking the Medical College Admission Test® (MCAT®) [7]. The prerequisite courses are specific to each medical school and typically include biology, English, chemistry, and organic chemistry but can also include social science, psychology, and calculus.* These courses provide a foundation in both scientific knowledge and critical thinking. After being accepted to medical school, the next phase begins.† Students spend the first one to two years of medical school learning the basic sciences of medicine, including normal physiology, anatomy, pathology, and pharmacology. Students are also introduced to clinical skills, learning how to take a medical history and perform a physical exam, as well as to the art of medicine, learning about the doctor–patient relationship, learning clinical approaches to patients, and forming a professional identity [8].

PROFESSIONAL IDENTITY

Professional identity, specifically the development of a professional identity, has long been thought about and discussed in medicine and medical education. A physician's professional identity has been defined by Cruess et al. as "a representation of self, achieved in stages over time during which the characteristics, values, and norms of the medical profession are internalized, resulting in an individual thinking, acting, and feeling like a physician" [8, p. 1447].

After developing a foundational knowledge base, students transition to primarily learning at the bedside and begin their clinical rotations, or clerkships, where they spend two-week to eight-week blocks of time with different medical specialties, including surgery, pediatrics, internal medicine, obstetrics and gynecology, family medicine, radiology, neurology, and psychiatry. In some programs, students undertake longitudinal clinical experiences, completing their required clinical experiences over the course of a year with portions of each week dedicated to the core medical specialties and continuity of patient care. During their time spent on clinical rotations, medical students

* Many individuals complete these prerequisite courses during their four-year degree; however, others complete them in post-baccalaureate programs.

† Medical school can be allopathic, resulting in a doctor of medicine degree, or osteopathic, resulting in a doctor of osteopathic medicine degree. Most medical schools are four-year programs; however, they can vary from three years or seven years depending on additional research years or time taken to acquire additional degrees.

gain further experience and skills in history taking, communication skills, working as a member of an interprofessional team, and developing diagnostic assessments and plans for patients. As they progress through medical school and enter their final year, medical students are entrusted with increasing levels of clinical responsibility. The "Acting Intern" or "Sub-Intern" rotation gives students the opportunity to serve in the role of the intern, or first-year physician, and places the student in the role of primary provider for their patients. The medical school education model is rooted in the acquisition of fundamental knowledge combined with the gradual development of clinical skills, professional identity, and level of responsibility over four years. During their time in medical school, students take standardized tests (USMLE® Step 1 and Step 2 or COMLEX-USA Level 1, Level 2 – Cognitive Evaluation, Level 2 – Performance Evaluation), which ensure that they have achieved a basic fund of knowledge and skill [9, 10]. During their final year in school, they apply and interview for residency programs in the specialties they are interested in and submit a ranked list of these programs [11, 12]. On Match Day, which is in March for the majority of residency programs, students find out in which residency program they will be starting in July – and residency programs learn who their next group of trainees will be [11].

FROM DOCTORATE TO LICENSED PHYSICIAN

Upon graduation from medical school with a degree in medicine,* each graduate is officially a doctor. However, a state medical license is required to practice clinically as a physician. While requirements vary by state, at a minimum, states generally require one year of post-graduate training and completion of an additional licensing exam (USMLE® Step 3 or COMLEX-USA Level 3) [13]. Becoming a board-certified physician, or a physician who is certified to practice in a particular field of medicine, requires completion of a residency which can range from three to seven years depending on the specialty and amount of time spent doing research [5]. The nature of training in residency varies dramatically based on the specialty. However, as a resident physician progresses through residency, they gradually gain increased autonomy. Every new physician completes an intern year during which they are supervised by an attending physician, or licensed and board-certified physician, as well as by upper-level residents. After completion of their intern year, the trainees assume additional levels of clinical and teaching responsibility while remaining under the supervision of an attending physician. The clinical experience can be a transformative process that allows not only for the acquisition of knowledge but also for the development of confidence in one's clinical judgment and decision-making, improvement in leadership skills, teaching skills, and increased ability to effectively work as a member of an interdisciplinary team [14].

The amount of knowledge available to medical students and physicians is vast and continually growing as new studies are published, new treatment protocols and recommendations are designed, and new pharmaceuticals are developed. Sources of information can range from standard textbooks to academic journals and online

* MD, DO, or MBBS for many international programs.

databases of peer-reviewed articles to online point-of-care medical resources. During residency, trainees learn to use their current clinical knowledge and clinical problem-solving not only to create a diagnostic assessment and treatment plan for a patient but also to identify their own knowledge gaps and the appropriate resource to address the clinical question. Because knowledge in this field is constantly evolving, students and trainees learn skills of lifelong knowledge acquisition and application.

Access to information provides a base to start from, but learning from patients, nurses, physical therapists, respiratory therapists, social workers, chaplains, attending physicians, and other members of an interprofessional team is a critical component of the development of a physician. Having easy access to a breadth of medical facts and empirical knowledge allows the focus of education in the hospital to be on clinical pearls and information that can't be easily turned into "fact" form.

The ability developed in residency of lifelong learning, and finding what works best for each individual, has been shaped by many factors, including the development of Free Open-Access Medical education, abbreviated FOAMed, originating in the field of emergency medicine in 2012 [15]. Physicians and other healthcare professionals worldwide share articles, clinical pearls, and videos on podcasts, blogs, and social media, particularly Twitter [15, 16]. Such information-sharing encourages ongoing self-directed learning and the development of online communities with active conversations about rising topics in medicine [15, 16].

ROLE MODELS AND MENTORSHIP IN MEDICINE

Throughout the process of becoming a physician, beginning as early as high school for some individuals, mentorship is an essential component [17]. From providing feedback on medical school and residency applications to guidance about navigating the unspoken messages in medicine about self-sacrifice and hiding emotions, mentors attend to and support personal and professional development [17]. For women in medicine, in particular, mentors can provide guidance about negotiating job contracts, professional development, dealing with sexual harassment and gender discrimination, and navigating practical logistics of balancing career and family – like how to continue breastfeeding while working 24-hour shifts [18]. Groups who have been marginalized in medicine, like those marginalized on the basis of race or ethnicity, also derive particular benefits from intentional mentoring relationships and programs [19]. In contrast to the one-on-one nature of mentorship, a role model is "a person whose behavior in a particular role is imitated by others" [20]. Role models can demonstrate what is possible in a career, model clinical skills to emulate, and are critical in professional formation for medical students, residents, and early-career physicians [18, 21]. Mentors can serve as both role models and mentors [18].

Mentors and role models of female physicians do not have to be exclusively female; however, it is helpful to have individuals with similar lived experiences to turn to, have an increased sense of community, and not feel alone [16, 18]. Further, there is a significant benefit in having someone who looks like you to demonstrate what is possible [22, 23]. While the number of female physicians continues to increase, women are not proportionately represented in higher-ranking positions in academic medicine, like Department Chairs and Full Professors, resulting in a lack of mentors for women pursuing and gaining seniority in academic medicine career paths [24, 25]. There are additional differences between specialties in the availability of female mentors as some specialties, like orthopedic surgery, remain male-dominated [26]. To increase the availability of mentorship to women and other groups who have been marginalized in medicine, many professional organizations have created specific mentorship programs that allow for connections outside of an individual's institution or organization [27, 28]. Connecting via online platforms, like Twitter, has been another avenue to foster connection and increase access to mentors and role models [16]. These mentoring opportunities serve a critical role in providing guidance for personal and professional development as well as networking opportunities [17].

BUILDING EXPERTISE IN MEDICINE

By the end of residency, graduates are expected to demonstrate competency in six core areas, as outlined by the ACGME: practice-based learning and improvement, patient care and procedural skills, systems-based practice, medical knowledge, interpersonal and communication skills, and professionalism [29]. The achievement of general and specialty-informed "milestones" in each of these core competencies is accomplished over the course of residency through a combination of knowledge acquisition, clinical experiences, and supervision with formative and summative feedback. Knowledge acquisition can occur in the form of learning from experts through lectures, textbooks, journal articles, teaching others, and research.

Clinical experiences vary by specialty but include rotating on various clinical services that fall within a resident's field of study. For surgical specialties, this includes spending time in the operating room and developing procedural skills. Through quality improvement initiatives and reflecting on challenging clinical scenarios, residents learn how to incorporate practice-based learning into their careers moving forward. Supervision and intentional feedback, by both upper-level residents and attending physicians, is a critical component of building expertise and developing into a competent physician. Through discussion of diagnostic reasoning and treatment decisions, the younger resident learns not only fundamental knowledge but also how to think and reason like a physician. Supervision and role modeling by other physicians fosters professional formation, including communication skills, working as a member of an interdisciplinary team, modeling how to think critically about patient care, and forming a professional identity [21].

One of the key skills developed during residency is the ability to sort through and interpret a large amount of medical knowledge and then apply it to a specific, individual patient and their particular clinical presentation. While the foundational information is readily available through external sources, practical pearls and the application of clinical knowledge are gained through clinical experience in residency and teaching by more senior physicians.

ONGOING INEQUITIES IN MEDICINE

While medicine has made great strides, and the number of female medical students is equal to that of male medical students, challenges remain and it would be remiss of us to not acknowledge areas in which progress is needed, particularly noting that they disproportionately affect groups who have been marginalized. Studies show that in addition to lower rates of promotion in academic medicine, there is disparity in pay between male and female physicians even after statistical adjustment for faculty rank, research time, and work hours [30–32]. Further, studies show that female faculty in academic medicine feel a lower sense of belonging in the workplace and increased incidence of gender discrimination and sexual harassment [32, 33]. Physicians from groups who have been marginalized, like those marginalized on the basis of race or ethnicity, face similar challenges including decreased feelings of belonging, workplace discrimination, and lower rates of promotion in academic medicine [33–35]. Women physicians in their early careers are also more likely to reduce work hours to part-time than men, which is thought to be due to work-family conflicts [36]. Women may also experience higher rates of burnout, depression, and career dissatisfaction than men [37].

Notably, there is significant gender segregation within medicine, with some specialties being predominantly male (i.e., orthopedic surgery, neurosurgery) and others being predominantly female (i.e., obstetrics/gynecology, pediatrics) [26, 38, 39]. These differences may relate to our discussion of gender schemas in Chapters 2 and 3, where women see themselves and are seen by others as being more suited for specialties that require warmth and nurturing rather than specialties that require technical skills [38].

Ultimately, while gender parity has been achieved in medical school, the development of early-career female physicians and their trajectory to becoming leaders in their fields of medicine looks very different from their male counterparts. Recommendations to improve gender equity have focused on the need to increase access to leadership opportunities and research funding, increase mentorship, improve work–life integration, and reduce implicit bias [40]. While many of these areas focus on the general culture of the workplace and remain important areas of focus, other areas – like policies related to parenting and childcare and work–life integration – may be more amenable in the future to a systems-based approach like Lean.

CONCLUSION

Regardless of the difficulties that remain for women in medicine, significant strides have been made, particularly when compared to the field of engineering. Bias, a discrepancy in access to research opportunities, and challenges related to mentorship still exist; however, access to fundamental knowledge is equal and promotes independence, confidence, and self-directed learning, which can reduce, although not eliminate, other factors involved.

REFERENCES

1. W. F. Bynum, *The History of Medicine: A Very Short Introduction*. Oxford: Oxford University Press, 2008.
2. University Archives & Records Center, "Brief Histories of the Schools of the University of Pennsylvania: School of Medicine," *University of Pennsylvania University Archives & Records Center*, 2022. https://archives.upenn.edu/exhibits/penn-history/school-histories/medicine (accessed Jan. 16, 2022).
3. K. M. Ludmerer, *Learning to Heal: The Development of American Medical Education*. Baltimore, MD: Johns Hopkins University Press, 1996.
4. K. M. Ludmerer, *Let Me Heal: The Opportunity to Preserve Excellence in American Medicine*. Oxford: Oxford University Press, 2015.
5. American Board of Medical Specialties, "Board Certification Requirements," American Board of Medical Specialties, 2022. https://www.abms.org/board-certification/board-certification-requirements/ (accessed Jan. 16, 2022).
6. Accreditation Council for Graduate Medical Education, "What We Do," Accreditation Council for Graduate Medical Education, 2022. https://www.acgme.org/what-we-do/overview/ (accessed Jan. 16, 2022).
7. Association of American Medical Colleges (AAMC), "Getting Into Medical School," *AAMC Students and Residents*, 2022. https://students-residents.aamc.org/choosing-medical-career/getting-medical-school (accessed Jan. 16, 2022).
8. R. L. Cruess, S. R. Cruess, J. D. Boudreau, L. Snell, and Y. Steinert, "Reframing Medical Education to Support Professional Identity Formation," *Academic Medicine*, vol. 89, no. 11, pp. 1446–1451, Nov. 2014, doi: 10.1097/ACM.0000000000000427.
9. American Osteopathic Association (AOA), "Prepping for COMLEX," American Osteopathic Association (AOA), 2022. https://osteopathic.org/students/preparing-for-comlex-exams/ (accessed Jan. 16, 2022).
10. United States Medical Licensing Examination (USMLE®), "Your USMLE Journey to Medical Licensure in the U.S.," United States Medical Licensing Examination (USMLE®), 2021. https://www.usmle.org/ (accessed Jan. 16, 2022).
11. American Osteopathic Association (AOA), "Navigating the Match," American Osteopathic Association (AOA), 2022. https://osteopathic.org/students/preparing-for-residency/match-guide/ (accessed Jan. 16, 2022).
12. Association of American Medical Colleges (AAMC), "How to Apply for Residency Positions," *AAMC Students and Residents*, 2022. https://students-residents.aamc.org/understanding-application-process/how-apply-residency-positions (accessed Jan. 16, 2022).
13. Federation of State Medical Boards (FSMB), "State Specific Requirements for Initial Medical Licensure," Federation of State Medical Boards (FSMB), 2018. https://www.fsmb.org/step-3/state-licensure/ (accessed Jan. 16, 2022).
14. C. E. Johnson, "The Transformative Process of Residency Education," *Academic Medicine*, vol. 75, no. 6, pp. 666–669, Jun. 2000, doi: 10.1097/00001888-200006000-00022.
15. O. Olusanya, J. Day, J. Kirk-Bayley, and T. Szakmany, "Free Open Access Med(ical edu)cation for Critical Care Practitioners," *Journal of the Intensive Care Society*, vol. 18, no. 1, pp. 2–7, Feb. 2017, doi: 10.1177/1751143716660726.
16. J. D. Lewis *et al.*, "Expanding Opportunities for Professional Development: Utilization of Twitter by Early Career Women in Academic Medicine and Science," *JMIR Medical Education*, vol. 4, no. 2, pp. e11140, Jul. 2018, doi: 10.2196/11140.
17. D. Sambunjak, S. E. Straus, and A. Marušić, "Mentoring in Academic Medicine: A Systematic Review," *JAMA*, vol. 296, no. 9, pp. 1103, Sep. 2006, doi: 10.1001/jama.296.9.1103.

18. J. L. Welch, H. L. Jimenez, J. Walthall, and S. E. Allen, "The Women in Emergency Medicine Mentoring Program: An Innovative Approach to Mentoring," *Journal of Graduate Medical Education*, vol. 4, no. 3, pp. 362–366, Sep. 2012, doi: 10.4300/JGME-D-11-00267.1.
19. E. Bonifacino, E. O. Ufomata, A. H. Farkas, R. Turner, and J. A. Corbelli, "Mentorship of Underrepresented Physicians and Trainees in Academic Medicine: a Systematic Review," *Journal of General Internal Medicine*, vol. 36, no. 4, pp. 1023–1034, Apr. 2021, doi: 10.1007/s11606-020-06478-7.
20. "Role Model," *Merriam-Webster.com Dictionary*. Accessed: Jan. 16, 2022. [Online]. Available: https://www.merriam-webster.com/dictionary/role%20model.
21. N. P. Kenny, K. V. Mann, and H. MacLeod, "Role Modeling in Physicians; Professional Formation: Reconsidering an Essential but Untapped Educational Strategy," *Academic Medicine*, vol. 78, no. 12, pp. 1203–1210, Dec. 2003, doi: 10.1097/00001888-200312000-00002.
22. The Lyda Hill Foundation & The Geena Davis Institute on Gender in Media, "Portray Her: Representations of Women STEM Characters in Media," The Lyda Hill Foundation & The Geena Davis Institute on Gender in Media, Online, c 2022. Accessed: Jan. 15, 2021. [Online]. Available: https://seejane.org/wp-content/uploads/portray-her-full-report.pdf.
23. S. Varalli, "Seeing is Believing: The Importance of Visible Role Models in Gender Equality," *The Art Of*, May 09, 2017. Accessed: Jan. 16, 2022. [Online]. Available: https://www.theartof.com/articles/seeing-is-believing-the-importance-of-visible-role-models-in-gender-equality.
24. K. P. Richter *et al.*, "Women Physicians and Promotion in Academic Medicine," *The New England Journal of Medicine*, vol. 383, no. 22, pp. 2148–2157, Nov. 2020, doi: 10.1056/NEJMsa1916935.
25. S. E. Nocco and A. R. Larson, "Promotion of Women Physicians in Academic Medicine," *Journal of Women's Health*, vol. 30, no. 6, pp. 864–871, Jun. 2021, doi: 10.1089/jwh.2019.7992.
26. B. Murphy, "These Medical Specialties Have the Biggest Gender Imbalances," *American Medical Association (AMA)*, Oct. 01, 2019. https://www.ama-assn.org/residents-students/specialty-profiles/these-medical-specialties-have-biggest-gender-imbalances (accessed Dec. 05, 2021).
27. American Medical Women's Association (AMWA), "AMWA Invests in You with Quality Mentorship," American Medical Women's Association (AMWA), 2022. https://www.amwa-doc.org/about-amwa/member-benefits-amwa/mentoring/ (accessed Jan. 16, 2022).
28. Ruth Jackson Orthopaedic Society, "About," *Ruth Jackson Orthopaedic Society*, 2019. http://www.rjos.org/index.php/about (accessed Jan. 16, 2022).
29. Stanford Medicine Graduate Medical Education, "Core Competencies," *Stanford Medicine*, 2022. https://med.stanford.edu/gme/housestaff/all-topics/core_competencies.html (accessed Jan. 16, 2022).
30. T. Wang, P. S. Douglas, and N. Reza, "Gender Gaps in Salary and Representation in Academic Internal Medicine Specialties in the US," *JAMA Internal Medicine*, vol. 181, no. 9, pp. 1255–1257, Sep. 2021, doi: 10.1001/jamainternmed.2021.3469.
31. R. Jagsi, K. A. Griffith, A. Stewart, D. Sambuco, R. DeCastro, and P. A. Ubel, "Gender Differences in Salary in a Recent Cohort of Early-Career Physician–Researchers," *Academic Medicine*, vol. 88, no. 11, pp. 1689–1699, Nov. 2013, doi: 10.1097/ACM.0b013e3182a71519.
32. N. B. Lyons *et al.*, "Gender Disparity Among American Medicine and Surgery Physicians: A Systematic Review," *The American Journal of the Medical Sciences*, vol. 361, no. 2, pp. 151–168, Feb. 2021, doi: 10.1016/j.amjms.2020.10.017.

33. L. H. Pololi, J. T. Civian, R. T. Brennan, A. L. Dottolo, and E. Krupat, "Experiencing the Culture of Academic Medicine: Gender Matters, A National Study," *Journal of General Internal Medicine*, vol. 28, no. 2, pp. 201–207, Feb. 2013, doi: 10.1007/s11606-012-2207-1.
34. D. Fang, "Racial and Ethnic Disparities in Faculty Promotion in Academic Medicine," *JAMA*, vol. 284, no. 9, pp. 1085, Sep. 2000, doi: 10.1001/jama.284.9.1085.
35. A. Filut, M. Alvarez, and M. Carnes, "Discrimination Toward Physicians of Color: A Systematic Review," *Journal of the National Medical Association*, vol. 112, no. 2, pp. 117–140, Apr. 2020, doi: 10.1016/j.jnma.2020.02.008.
36. E. Frank, Z. Zhao, S. Sen, and C. Guille, "Gender Disparities in Work and Parental Status Among Early Career Physicians," *JAMA Network Open*, vol. 2, no. 8, pp. e198340, Aug. 2019, doi: 10.1001/jamanetworkopen.2019.8340.
37. K. Templeton *et al.*, "Gender-Based Differences in Burnout: Issues Faced by Women Physicians," *NAM Perspect*, May 2019, doi: 10.31478/201905a.
38. E. Pelley, and M. Carnes, "When a Specialty Becomes 'Women's Work': Trends in and Implications of Specialty Gender Segregation in Medicine," *Academic Medicine*, vol. 95, no. 10, pp. 1499–1506, Oct. 2020, doi: 10.1097/ACM.0000000000003555.
39. C. C. Chambers, S. B. Ihnow, E. J. Monroe, and L. I. Suleiman, "Women in Orthopaedic Surgery: Population Trends in Trainees and Practicing Surgeons," *Journal of Bone and Joint Surgery*, vol. 100, no. 17, pp. e116(1-7), Sep. 2018, doi: 10.2106/JBJS.17.01291.
40. A. Westring, J. M. McDonald, P. Carr, and J. A. Grisso, "An Integrated Framework for Gender Equity in Academic Medicine," *Academic Medicine*, vol. 91, no. 8, pp. 1041–1044, Aug. 2016, doi: 10.1097/ACM.0000000000001275.

9 An Engineer's Learning Environment

In Chapter 7, we wrote at length about Toyota's success in the United States and world automotive markets. Additionally, we wrote of the design of the manufacturing and engineering system that produced those results and the methods for their success. As we now turn to the specific efforts of enabling the success of engineers, we note the words of Durward Sobek II:

> Toyota does not leave the development of engineering skill and expertise to chance. It very conscientiously and systematically develops its engineers to ensure that the product development organization has the skills and expertise needed to stay at the forefront of technology.
>
> **[1, p. 249]**

Durward wrote this in 1997. In the year 2021, compared to other automotive companies, Toyota is Number 1 in unit volume and brand recognition [2, 3]. Clearly, Toyota's methods have been effective from a business standpoint. We now shift to examining the current approach to work within non-Lean firms in the United States.

A TYPICAL, BUT NOT DESIRABLE, EXAMPLE OF PRODUCT DEVELOPMENT

Let's look at a hypothetical example of the development approach within what we will call a traditional product development organization.

PRODUCT DEVELOPMENT IN THE TRADITIONAL WAY

A new-to-career female engineer has been working on a design solution for an important product feature for a number of weeks. While working in a progressive organization, she has received very good support from people who she has directly worked with and has created a solid technical approach to the work. As part of her work, she is now the "expert" for this area of design as no one else has the level of detailed understanding that she has established even if the overall scope (breadth of the work) may not be large. Following established design practices, she is asked to put together a design review and present it to a broader set of engineers.

As part of the normal way work is done, and modeling those with significant experience, she creates a presentation slide deck that communicates the

DOI: 10.1201/9781003205814-14

problem, considerations, and solution for the design review. However, even though well done, these slides are largely administrative and lack technical detail because the technical details are normally "talked through" and in some cases the designer is expected to verbally "defend" their work.

Since this is a hardware solution, she will be well prepared to "spin CAD" (computer-aided design) as part of the review, which means that she can show the reviewers detailed design elements on the screen in real time. In terms of the design process, she has developed her CAD design to an "almost done state." She has done this so the reviewers can see the maturity of her design solution. Additionally, in the review, she will largely focus on the single solution, identified as the best solution, that she and others helped find. As we will see later in this chapter, from a design process she is at the "design" phase of "best idea" and preparing to move to the "build" phase. Because the CAD is basically done, any changes at this point will more than likely impact the schedule.

Now that the engineer has prepared for the review, let's step into the review meeting.

THE TRADITIONAL DESIGN REVIEW

In the design review, which includes individuals who are new to the problem, the engineer starts by presenting her slides on the overhead projector. She opens the review with the problem, general background, and summary slides which generate some discussion. However, the team is really there to review her design. So, she replaces the slides on the screen with the CAD to present the design detail of her solution. She also has sample parts to show. As the team looks at the CAD and passes the parts around the table, they begin to ask insightful questions about the specific design. However, even with very supportive team members, the meeting eventually migrates into brainstorming design considerations that she and those she has worked with have already considered. The other ideas suggested weren't looked at because of constraints or knowledge that she did not know existed. No one is actually presenting data to support these ideas; they are just voicing their technical insights based upon their experiences which in some cases are technically helpful but in other cases are mixed with bias and opinion. As the more senior engineers offer up their own experiences and expertise, the engineer and group begin to question her fundamental design approach and urge reevaluation of other design choices.

Even with the setback of questioning the design, the team feels good about the review because of the positive work climate and tells the engineer that she did a "good job." However, since there is still more work to do at this step, she has been asked to set up another meeting to review the new information before agreeing on the final design. At the end of the meeting, the new-to-career engineer will take her notes and slides (which she has been asked to update and then distribute) and work through the information that she learned. (We use the word information rather than knowledge because it is unclear how much knowledge she gained from the review.)

Even though she has to redo part of her work, the team views her design efforts as "A" level work and that what she learned in the meeting is a part of the process. How did she get an "A" when she had to redo part of her work? As we will see later in the chapter, the entire system is centered around what is termed Build–Test–Fix. This may be ok for the new-to-career male engineer, maybe even considered a "rite of passage." However, for her, as the "lone" woman in the room, given what we have learned in Chapters 2 and 3 about the tendency to view women as less competent than men, the results of the review may result in the other engineers unconsciously viewing her as less competent. She, in turn, may feel less competent and have less confidence in her ability to be successful.

Although this example is specific to designing hardware, a physical object that we can hold in our hands, we can create a version of this story for virtually any problem that an engineer is being asked to solve. We can imagine the story being about an engineer working to create a new customer support process in which the majority of the review is centered around a flow chart and data linkages to computer systems with the creation of complex data analysis tools. We can imagine a version in which an engineer is creating a new manufacturing process that will be put into a complex factory setting in which the sequencing of the process installation and validation of its performance is the primary challenge. In any of these cases, they likely follow the same basic flow of work: understand your objective, find your best solution, present it to others, and see how it goes.

The importance of this story is to get back to the fundamental piece that can be controlled, which is how the work is done. However, before we proceed in that direction, let's look at how someone achieved that first step in becoming an engineer, meaning, a degree in hand.

HOW ENGINEERS GET THEIR DEGREES

Obtaining an engineering degree is accomplished through a very structured approach that focuses on a few key aspects:

1) Gaining the technical knowledge in your area of focus
2) Learning how to solve problems
3) Understanding how to make a contribution to a broader purpose

In the last term of calculus, on the first day of class, the instructor for Bob's class told the students, "For you that are in engineering, you will never use what we will cover in this class, the sole purpose of this class for you is to teach you how to think." It is more than likely that some engineers did go on to use this content, but Bob certainly didn't. To achieve a degree in engineering, there is a list of the required courses for the chosen discipline. Then, for each specific term, the student signed up for the needed classes, bought the required books, and went to class. At the beginning of each course, the hopeful future engineer focused on learning the required knowledge to achieve success on that first test. The combination of the amount of effort put in for that first test and the person's innate ability leads to their level of success. Throughout the remainder of the term, the student continued to learn, moving along each step of the way toward the end of the term. At the end of the term, the grade that was earned demonstrated how much learning occurred and knowledge acquired.

Throughout the term, the learning process's driving force was access to available knowledge, a person's effort, general capability, and, in many cases, an individual's passion for the topic. In general, this structured learning process is what enabled men and women to have a more level playing field, as shown in Figure 3.1. However, as with many situations with inherent bias, the learning structure alone doesn't make up the whole experience and other climate factors may negatively affect women students. However, the structured learning process and access to a high level of available knowledge, controlled by the student, can help to increase confidence and mitigate bias. Much like we have talked about the progressive firm, we know that some universities are more progressive in regard to women in STEM than others.

HOW THE LEARNING OCCURS AS A NEW ENGINEER

For the new-to-career engineer, this learning process that we saw in school likely takes a dramatic turn when someone enters the workforce. During their first few years working in engineering, they will more than likely lack access to structured knowledge and structured learning. They are typically dependent on gathering the necessary knowledge from those around them and are therefore not in control of their learning. As previously covered, in some areas of development, written knowledge accessible to an engineer might make up only 20% of the knowledge needed to do the work. The remaining knowledge comes from peers and, in the worst situations, from recreating knowledge that the organization has lost due to the departure of senior-level engineers. This learning could be related to the specific technical work for an engineer's design area or organizational processes such as drawing releases, submitting test requests, or finding customer data.

At this point, the overall work climate and nature of interpersonal interactions come into play. If engineers feel they will be perceived as incompetent if they ask too many questions, they may avoid asking for help even if another engineer has a readily available answer. Further, having to ask for help on a continual basis because there is no other way to easily find the answers can lead to feelings of inadequacy and low job satisfaction. These issues exist for both men and women in engineering. However, given our underlying preconceptions about men and women, there is a greater risk for a woman engineer that she will be perceived as less competent if she has to seek help. This can have many downstream effects including her perceptions of her own abilities, the tasks she is assigned, her overall job satisfaction, and ultimately, her desire to remain in the field of engineering. Without access to knowledge and, therefore, the ability to effectively do her job, she lacks a significant amount of control over her career.

COMPARISON TO A PHYSICIAN IN RESIDENCY

As we saw in Chapter 8, in addition to pre-med students' efforts to sustain a high GPA (Grade Point Average), they are also working toward high marks on the critical MCAT (Medical College Admission Test®). In combination with the college GPA and extracurricular activities, this test score is one of the factors contributing to the applicant being one of the roughly 40% of applicants accepted to medical school [4]. When a student enters medical school, she or he enters another form of structured learning. So, even if the student received her bachelor's degree in a male-dominated program such as engineering or physics, she enters into an educational program that is now 50% women. Like in engineering, there is a portion of medical school during which students are focused on learning the required information through lectures and access to knowledge in textbooks with a series of exams to test their knowledge and build their confidence. They also spend a significant portion of their time gaining direct, hands-on clinical experiences through a structured approach – clinical rotations. While issues of bias still arise through subjective grading processes associated with clinical rotations and day-to-day experiences, students still have a portion of control over their learning.

Once in a residency program, a typical physician is able to gain a significant amount of knowledge through lectures and access to reusable knowledge. This information is enhanced and shaped through clinical encounters with patients and direct feedback and supervision by an attending physician. As patients, we want to know that health problems are being fully thought through; however, the way the care is delivered also matters. Having a structured learning process allows for both the ongoing development of knowledge to provide the best care possible and the learning of the human side of medicine.

HOW THE WORK IS DONE AS A NEW-TO-CAREER ENGINEER

Since we have explored how the lack of available knowledge impacts the work environment, let's now move to the area of "how the work is done" and how this drives the learning process. As we saw in the story above, a common learning method used in product development is the design process of Build–Test–Fix. A more technical description for this is "point-based design." This development approach largely depends on the level of initial knowledge the individual or team has prior to starting the design work. As shown in Figure 9.1, the work begins with the team understanding the overall objectives and requirements and identifying possible solutions. As the team evaluates each of the possible solutions, they may have a team dynamic in

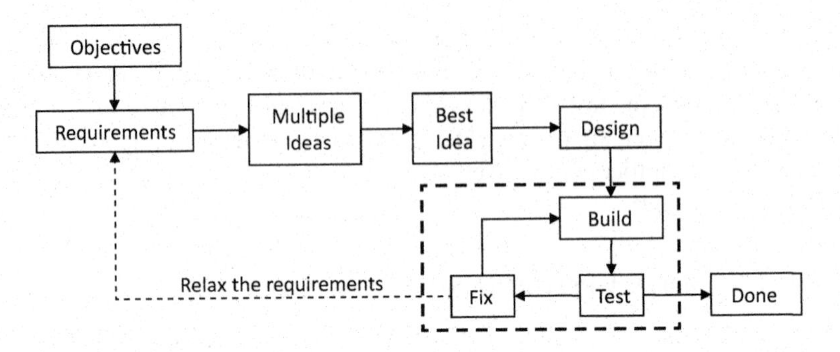

FIGURE 9.1 Traditional development – Build–Test–Fix – point-based design.

which the individual with the loudest voice or the most confidence determines the solutions considered and the final selection. For those in the minority on the team, including women on the team, this situation may be disempowering and may have many downstream effects including reducing their feelings of contributing to the business and decreasing confidence. Further, listening to the loudest voice likely doesn't result in choosing the best solution. Nevertheless, the team has arrived at what they feel is the best idea.

With the best idea chosen, the team then creates a design and begins to work through the Build–Test–Fix cycle with the goal of meeting the defined requirements. In a situation with a significant lack of initial knowledge, it can be challenging to achieve success with the first design choice. Additionally, in some cases, before the team or individual can achieve a solution that meets the requirements, they may run out of money, time, or resources and are forced to ship the last best solution they have. This action is further aggravated by a sense of pride the team may have developed for their best idea and a belief that if anyone can figure out how to make the idea work, they can – after all, it is the best idea. At some point though, there is not enough time to move to an alternate solution – independent of money and resources. In what might be the worst situation, the team may be forced to relook at the requirements with the required fix effort driving that discussion. This discussion may start with "if we can relax this requirement, we have a design we can deliver."

From our causal diagram (Figure 6.2), the method of Build–Test–Fix is at the center of "Poor Work Methods." Build–Test–Fix in a team situation can create an environment in which the loudest voice can dominate the development work and disempower individuals on the team.

As we saw in our story above, the design process loop of Build–Test–Fix can have negative outcomes on an individual's confidence, job satisfaction, and perceived competency by others given the inherent gender schemas we all have. In some cases, the selected design approach works great the first time out and everyone is happy. The immediate success may increase the confidence of the new female engineer.

However, if the organization attributes the success of the design to the task being simple or to luck rather than to the engineer's abilities, as people tend to do for women's successes, then whatever perceptions of the engineer's abilities that already exist remain entrenched. In most cases though, having to go through multiple loops of Build–Test–Fix can hardly be considered a confidence builder. The most challenging aspect for a woman engineer is that, unlike a male peer who can fade into the background among his other male peers, when challenges exist within this development method her work is very visible.

THE DESIGN PROCESS, A THIRD-GRADE EXPERIENCE

Bob was asked to come into a third-grade classroom, taught by his daughter, Jessica, to be the Engineer Judge for a class STEM project. The students were tasked with designing a candy launcher using household items such as paper clips, rubber bands, popsicle sticks, and tape. When he walked into the class, Jessica was at the front of the room just starting to present a series of slides projected on the screen. She began with the objective of the project and what the students were trying to build. She then proceeded to talk about their role as an engineer in doing this work. She talked about the materials they had. She talked about the importance of them working together as a team and listening to their teammate because that would make the job easier. As a dad, and an engineer, Bob was proud of how well she communicated the expectations and the engineering problem – clear, concise, and direct.

Then, Jessica brought up a slide that gave him pause about the reality of the current world of engineering. The slide was very well done, with a clear description of the design process that these third graders would use. That design process was Build–Test–Fix. The flowchart (yes, a flowchart) specifically communicated the "loopback" that was expected for their work. It was exactly what these third graders should be using; however, the realization that a significant number of engineering organizations are using the same approach as what third graders are being taught was tough to hear as she said to the class, "this is how engineers do their job."

After Jessica finished her instructions, the class was put into teams of two to go off and solve their engineering problem. By the end of the allotted time, all teams were able to go through enough Build–Test–Fix cycles in order to build and then demonstrate that the design of their candy launcher met the requirements. Everyone had fun.

In Chapter 11, we will cover a development method and principle that replaces Build–Test–Fix and puts a design team or engineer in much more control of their learning. In Lean Development, it is called set-based design or set-based concurrent engineering (SBCE). This method was promoted as a fundamentally better way of design in a 1999 Sloan Management Review article by Durward Sobek II, Allen

Ward, and Jeffery Liker based on research of Toyota [5]. Set-Based Design promotes a learning environment and can address the inherent bias that can come out with the method of Build–Test–Fix.

One final thought on Build–Test–Fix, which may be one more than is needed.

WHAT THE INTERNET TELLS US OF BUILD–TEST–FIX

If you conduct an internet search of "STEM Engineering Design Process Posters," you will get dozens of posters that communicate the design process of Build–Test–Fix, all in colorful and engaging images designed for elementary school-age students. (Though they may generally replace the word fix with improving.)

It is great that students as young as second or third graders are being taught a direct method of engineering. Without this overt method of teaching, most students would have little understanding of how an engineer approaches the work.

CHANGING WHAT CAN BE CHANGED

In the first part of this book, we described some of the challenges facing women, particularly those working in male-dominated environments. These include perceptions of competence and ability, having a supportive work environment, feelings of job satisfaction, and the nature of the interactions with others. Over the course of this book, we have shifted our focus from the individual to a systems-based perspective and, using our causal diagram, identified specific root causes – changing the way the work is done, increasing the availability of knowledge, and developing of role models.

We know that we can't entirely control the work climate as it is shaped by how people behave, but we can control the way the work is done and whether a firm is progressive or not. In Figure 9.2, we propose that if we add good work methods into the development process (using the methods of Lean which creates value, the *x*-axis), we can improve the climate (the *y*-axis). Additionally, if we move an organization from non-progressive to progressive, we improve the climate.

However, before we further break down the structure of Figure 9.2, let's take a simple example that may be closer to home for some of us.

LET'S LOOK AT FIGURE 9.2, LIKE A SOCCER PARENT

In the game of soccer, whether it is a team of 8-year-olds or 16-year-olds, the basic work structure is the same. On the *x*-axis, the game's structure, let's assume there are two inputs: the rules of the game and each player playing their

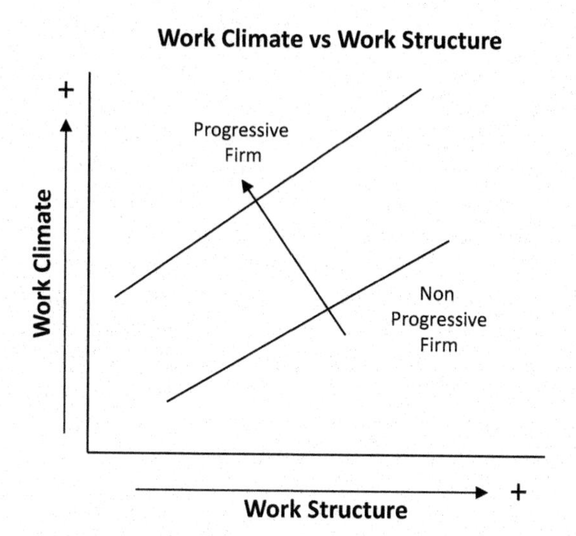

FIGURE 9.2 Work structure driving climate.

position given the execution of a play. These two inputs establish how someone is supposed to act on the field and what they can expect from their teammates. When thinking about team climate, this structure is what may finally address the behavior of the continual "ball hog." It might also bring out the amazing talent in a team that unfortunately was covered up by the "ball hog."

The location of the diagonal lines in relation to the *y*-axis represents the role of a coach and the tone he or she sets for the team, which can range from autocratic to democratic. These lines are determined by the coaching style during practice and games and influence how players treat each other. The difference between these lines can drive players (or the parents of players) from one coaching style or team to another team. This change in coaching style may be the change that brings out the best skills and capabilities of all team members, which were previously covered up by ineffective teamwork. (No comment on how parents can negatively influence this.)

Returning back to our engineering work, using Figure 9.2 and thinking specifically about access to available knowledge, we can rationalize where people sit on the *x*-axis. For example, suppose a female engineer is dependent on her peers for a majority of her learning, and each of those interactions is less than positive experience (e.g., she is working in a non-progressive firm). In that case, we could assume she sits both on the left side of the *x*-axis and the lower diagonal line, as she is experiencing a poor work climate. However, if by changing the structure of the work providing access to available knowledge, we could dramatically change the nature of her interactions with her colleagues we could then move her point on the *x*-axis to the right. Based on this graph, we would expect the work climate that she is experiencing to improve. This does nothing to move the organization to progressive, it literally just changes the nature of the personal interactions and reduces potential negative interactions. In this case, we are specifically discussing interactions where her ability to effectively do her job is reliant on the willingness of others to help. Let's compare this with training in medicine. The new physician automatically starts further to the right on the x-axis simply because of the way the work is done and her access to reusable knowledge. She may still practice in a work environment similar to a non-progressive firm and have to deal with negative interactions from others including people doubting that she is indeed a physician; however, her ability to learn and to make clinical decisions is not entirely reliant on others. While the physician is still working in a non-progressive climate with challenges remaining in terms of being heard and listened to, the way the work is done does improve the physician's overall experience at work in comparison to a female engineer operating in a similar environment.

In another example, if our early-to-career engineer has access to knowledge, we might see the interactions in which she is asking for assistance go from daily to once a week. This reduction in potential negative interactions is positive for her. In addition, it is possible that her peers and supervisors may develop a different perspective of her capability (e.g., lower bias) when she is no longer needing to seek out such frequent help. The weekly discussions may take on a different tone, perhaps even positive, and may shift from simple provision of knowledge to collaboration and innovation. This could be measured by the woman who has regular interactions with experienced engineers, "I used to have 5 negative experiences a week and now I have one positive experience a week." Additionally, the organization's overall productivity increases, as less time is spent seeking out an answer to a question when the engineer has easy access to the necessary information, time can instead be spent discussing new ideas and solutions.

In Figure 9.2, we have one line representing a progressive firm and another line representing a non-progressive firm. These exist on a continuum and can shift up and down the y-axis. On our causal diagram (Figure 6.2), we listed six components of the work environment that influence Work Climate: Work–Life Balance, Being Valued, Being Respected, Acceptance of Diversity, Behaviors, and Role Models. In considering a firm on the spectrum of non-progressive to progressive, we could create a quantitative assessment for each of the six categories and sum their total. By creating some defined criteria for each of them we could come up with a scale of 1 to 5. In our example above, in which the woman engineer was having multiple negative interactions a week which she might put under the Being Respected category, she may score that part of her work as a two. However, now that she is having one positive experience a week and fewer negative interactions, she may change the Being Respected category score to 4. In Chapter 14, when we work through the creation of development plans, we will see that this exercise of having an individual score each of the six factors that influence work climate can be used to find where an organization needs to make changes to improve the work climate.

Of note, for an early-to-career male engineer in exactly the same situation, daily interactions with experienced engineers may be considered positive by all involved.

In the context of our gender schema, the male engineer is more likely to be viewed as someone who is taking initiative and being an engaged member of the team rather than someone lacking in competence. Further, the experienced engineers may look forward to the discussions because that early-to-career engineer is just like them at the start of their careers, and if they talk just before lunch, they may have lunch together or head out to the company basketball court and shoot some hoops.

CONCLUSION

There is a basic tenant of any process improvement effort: you can't improve what you don't understand. The objective of Chapter 9 was to create a common understanding of the current learning environment of an engineer and then understand which components should be changed. We identified two key components of training in engineering: lack of access to knowledge and the learning process of Build–Test–Fix. Each of these factors increases the negative effects of bias and our inherent gender schema and can reduce overall work productivity and outcomes. While organizationally both of these could be lumped under the broad category of organizational costs or just the way of doing business, they have broad ramifications – affecting what the organization is trying to accomplish and shaping the overall work climate. The lack of access to knowledge and the learning process of Build–Test–Fix are therefore two key targets for a system-level change.

REFERENCES

1. D. K. Sobek II, *Principles that Shape Product Development Systems: A Toyota-Chrysler Comparison.* Ann Arbor, MI: University of Michigan, 1997.
2. Focus2Move, "Global Auto Market 2021. Ford (-0.4%) is the only group to report losses in Q1," *Focus2Move*, Apr. 26, 2021. https://www.focus2move.com/world-car-group-ranking/ (accessed Jan. 13, 2022).
3. Brand Finance, "Automotive Industry 2021 Ranking," *Brand Finance Brandirectory.* https://brandirectory.com/rankings/auto/table.
4. Association of American Medical Colleges (AAMC), "2019 Fall Applicant, Matriculant, and Enrollment Data Tables," Association of American Medical Colleges (AAMC), Dec. 2019. Accessed: Jul. 05, 2021. [Online]. Available: https://www.aamc.org/media/38821/download.
5. D. K. Sobek II, A. C. Ward, and J. K. Liker, "Toyota's Principles of Set-Based Concurrent Engineering," *Sloan Management Review*, vol. 40, no. 2, pp. 67–83, Winter 1999.

Part III

Summary

In Part III, we brought in three elements of our problem-solving effort.

- In Chapter 7, we gave a brief overview of the history of lean (the elimination of waste) through our learnings from the North American automotive industry during the early 1980s. During the 1980s, lean then moved into virtually every US industry. We then introduced Lean Development, which at its core is focused on increasing the rate of learning.
- In Chapter 8, we outlined the training and learning approach of our case study, a physician in residency. At the core of the approach is work in a designed system of learning.
- In Chapter 9, we described how the typical new-to-career engineer is trained, how they learn, and how they approach their development work through build–test–fix methods. To address the issues associated with this development method, we introduced the method of set-based design.

As we move into the implementation of Lean Development to address the three root causes from Chapter 6, we will first focus on addressing two of the overarching areas: Control Over Career and Creating Satisfying Work. Improving these two areas will naturally improve the Work Climate. Based on our causal diagram (Figure 6.2), addressing the root cause of Reusable Knowledge increases Confidence, which then reduces Bias. In Parts IV and V, we will find that most other areas within the Opportunity box of the causal diagram either directly or indirectly influence Work Climate, including an individual's Technical Depth or the presence of Technical

DOI: 10.1201/9781003205814-15

Coaching. Using Lean Development as the foundation, this brings us to two strategies to address the issues founds in the causal diagram (Figure 6.2):

STRATEGIES:

1. Empowering control over career (Part IV)
2. Enabling leaders to lead – creating satisfying work (Part V)

Using these two strategies, we will see the fundamental business purpose, as measured through increasing profit and revenue achieved. To be clear though, the nuance of our use of Lean Development is overtly focused on the major element that delivers these business results, the people doing the work that customers value. The profit and revenue will follow from that effort.

Part IV

Strategy 1

Empowering Control over Career

10 Building a Level Playing Field

In Chapter 9, we saw in our hypothetical example, a new-to-career engineer getting her first exposure to the way that the development and design review processes are managed in, what we will call, a typical progressive firm. She saw the positive aspects of a progressive firm – feeling supported by her peers during the review process. However, she still had to go back and redo her work because she lacked critical information and access to other perspectives. Ultimately, though, the most fundamental thing she lacked was a development process that was grounded in learning and gave her control over her learning.

We can summarize attributes of a typical product development process to be:

1. Phase gate reviews: These are organizationally defined reviews that either allow work to move forward or hold it up. In some cases, they may prevent an entire program from moving to the next step. These reviews may be very administrative in nature.
2. Built–Test–Fix
3. Separation of the business side (the management team) from the technical side (the engineering team)
4. And, ultimately, depending on schedule pressures, the "Just Get It Done" mentality

In our example, the new engineer had her first exposure to Phase Gate Review (hold a design review) and Build–Test–Fix (take the new information and redo your work).

In Part IV, we shift from the typical product development approach to a Lean Development based approach. Specifically, focusing on how Lean Development can improve one of the three major contributing factors to a low retention rate of women in engineering, an individual engineer's control over their career, by targeting two of our root causes, access to reusable knowledge, and access to role models. Before applying a solution to a problem, it is necessary to fully understand both the problem and the principles underlying the solution to the problem. Chapter 10 therefore further details some of the challenges that women in engineering face in gaining control over their career, prior solutions that have been tried, and a Lean Development based solution – the A3 report – which increases access to reusable knowledge. Chapter 11 delineates the building blocks of Lean Development while Chapter 12 explores the importance of role models and technical coaches.

DOI: 10.1201/9781003205814-17

THE UNLEVELED PLAYING FIELD FOR WOMEN

When the available knowledge is not in a reusable form and the vast majority of learning is attained from others, the learning process can feel more like playing 20 questions. If you happen to ask the right question, the answer to the entire problem may appear obvious. However, you could just as easily ask a series of "wrong" questions leading to increased frustration and delaying the entire process. In an environment where there are implicit biases and differences in levels of confidence, as we explored in Chapter 2, small advantages and disadvantages add up over time. One could imagine a scenario in which two engineers, one man and one woman, asked their manager a series of questions. Given that the manager, regardless of gender, likely has an unconscious belief that men are more competent than women, they are more likely to view the man asking questions favorably (i.e., taking initiative, dedicated to their work) and are more likely to view the woman unfavorably for asking questions (i.e., not as knowledgeable, not independent) [1].

The difference may be very small – the manager may view the male engineer in only the smallest more positive light and the female engineer in only the smallest more negative light. However, over the course of their time working at the engineering firm, these small interactions continue to add up over time. In contrast, a system that rewards the process of learning as well as facilitates access to a fundamental base of knowledge could both shift the focus from outcomes to the process of the work itself and reduce the need to ask a series of questions to obtain information. While this change won't remove bias from the workplace or resolve differences in confidence overnight, it can reduce the number of advantages (or disadvantages) that are accumulated with the goal of leveling the playing field.

THE WORK OF RELATIONAL PRACTICE

In Chapter 5, we wrote of the relational job responsibilities that women may be drawn to or in some cases encouraged to move to. This stems from our underlying gender schema, associating men with assertiveness and competency and, therefore, perceived as being naturally more suited for technical-based occupations, and women being warm and nurturing, or being naturally more suited for relationship-based roles [1, 2]. In engineering, we see this in men typically being tasked with, and based within, product development, with their work technically focused and relational practices de-emphasized [3]. In contrast, women are shifted to roles that focus on relational practices, like technical planning and supporting a cohesive work team, rather than technical roles [3]. This creates a separation of responsibilities.

As described in Joyce Fletcher's *Disappearing Acts: Gender, Power, and Relational Practice at Work,* there are four types of relational practices:

- "*Preserving*: Preserving the project through task accomplishment;
- *Mutual Empowerment*: Empowering others to enhance project effectiveness;
- *Self-Achieving*: Empowering self to achieve project goals; and
- *Creating Team*: Creating and sustaining group life in the service of project goals."

[2, p. 48]

These relational practices can occur in many areas of engineering – from facilitating team meetings to ensuring that a design makes it from the design phase to production.

If a male engineer created a new design, it could be reasonable to extend his responsibility to include moving that design through the development process and into the new product, requiring relational practices like negotiating with support organizations which may have challenging interpersonal situations. However, the development system is not designed for the process to happen this way. In a typical situation, the engineer hands his work off to an engineer working at the program level. When relational roles, like coordinating the execution of a project or ensuring that each member of the team is able to accomplish their tasks, become full-time work for women, it limits their ability to increase their technical depth and have balanced work; two key areas of opportunity for increasing retention of women in engineering (see Figure 6.2).

For the organization, while it may not value relational practices as highly as technical work, it usually understands the need. The organization may believe that women will do a better job than men in roles requiring the use of relational practices. From the organization's perspective, shifting women into relational roles may be seen as a good use of resources. However, this practice not only creates further challenges for women in the workplace, but it is also based on an unconscious belief about women that for a particular individual woman may or may not be true. Further, it may not be a job that a particular woman engineer finds professionally satisfying or fulfilling, feels that she is competent at or suited for, or simply does not want. So, how do we want to approach this? We approach it as another element of "how the work is done." In this case, we are not focusing on the method of the work but what the work is – looking at the division of responsibilities within the work. Understanding this dynamic within an organization is critical as we move to broaden our design of the system of engineering.

It is important to note, though, that expecting men pick up the relational practices for their own work or the needs of the broader team without providing training and coaching on how to do that is not fair to anyone and may be detrimental to the organization. We will work through how to approach this challenge later in the chapter.

CHANGE THE WAY THE WORK IS DONE – FINDING POSSIBLE SOLUTIONS

Before we discuss Lean Development as the central method to change what the work is and how it is done, let's look at a few other business management strategies and methods of doing the work for context.

Management by Objective

Throughout the book, we have separated out two clear aspects of retention: how people interact in the act of doing their jobs and how the work is done. In the case of a new-to-career engineer, how the work is done has a profound impact on how they are trained to achieve the end result. A classic methodology identified by Peter F. Drucker, in his 1954 book *The Practice of Management* wrote "To manage the business means, therefore, to *manage by objectives*" [4, p. 12]. For progressive firms in the 1980s, this was termed "management by objective," or MBO, and was applied at the individual level. This freedom of an individual to establish how they did the work to achieve the objective they had been asked to accomplish was viewed as both a positive movement and a recruiting tool for new engineers to a firm. With the objective set, it was up to the individual to establish how they got there. As an engineer, what could be better than that?

For many work environments, the practice of MBO and the freedom people had in how the work was done was a positive change over command and control. However, as an organization grows, the freedom of how the work is done may become constrained. As much as we want to envision that MBO allows for an endless array of "how's," eventually organizations settle on a common path and everyone is expected to follow the "how" that is considered to be the norm. This common "how" may be less about what has been proven as the best management method and more about the personalities of the people who make up the organization and the power they hold. Additionally, just telling someone the objective doesn't mean you will draw the best results out of an individual or that the organization will see the best outcome. Some people may thrive in an MBO environment while others may not.

Reengineering – "Starting over"

The concept of "reengineering" is clearly linked to the 1993 book, *Reengineering the Corporation*. The authors' quick definition of reengineering is "starting over" rather than tinkering to get incremental improvements [5]. Reengineering starts with the idea that many activities within a corporation are managed as a series of sequential steps that have been broken down into simple tasks. Reengineering looks at those tasks as a process, with a beginning and an end, with the goal of changing the process to deliver radically improved business performance. Through the work of reengineering, tasks may be combined, run in parallel, or eliminated. While just improving the existing series of tasks may yield 10% increased performance, reengineering expects an improvement of 50% or better. Of note, within the book, the authors highlight the success Walmart was having, in 1993, in its use of reengineering its business processes, compared to Sears. A correct call, 30 years later.

The challenge of reengineering as an approach for our work, retaining women engineers, is that it is not driven by fundamental principles that can be broadly applied. However, the key elements of "starting over" and looking at the entire process are elements that we are using in our effort. Recognizing how the work is currently done is the first step in changing it.

Deductive or Inductive Thinking – Approaching a Problem

Part of changing how the work is done is to understand how we approach problems. A deductive thinking approach is to identify a problem, define the problem, establish possible solutions, and implement a solution. In contrast, inductive thinking is characterized as "the ability to first recognize a powerful solution and then seek the problems it might solve, problems the company probably doesn't even know it has" [5, p. 84]. Inductive thinking is basically looking at the problem from the bottom up compared to deductive thinking which looks at the problem from the top down.

Deductive thinking or problem-solving works well when a business or organization can meet its needs with incremental changes. However, when the business needs large changes, because of either what is known now or what is assumed to come in the future, only inductive thinking will support those goals.

For our situation, specifically women only making up 15% of engineers in the field and the lack of significant improvement over time [6], we need to have an approach to address the problem based on inductive thinking. Specifically, what tools or methods do we have access to or need to create in order to cause a dramatic shift? An inductive approach would be to start by recognizing that within work situations, there exists the possibility of bias which can impact many areas. The next step would be to look for tools or methods available to reduce the impact of bias within the workplace, decreasing both the frequency that someone experiences bias and the magnitude of the bias they experience. In looking for powerful system-based solutions to reduce the impact of bias in the workplace, we turn to Lean Development.

CHANGE THE WAY THE WORK IS DONE – LEAN DEVELOPMENT

Lean – The Elimination of Waste

In Chapter 7, we saw that the economic pressures on the US automotive companies in the 1980s initiated a shift to Lean Production. Exploration of this method in the 1990s provided clarity about "the how" to do the work. In James Womack and Daniel T. Jones's 1996 book, *Lean Thinking*, they describe lean as the elimination of waste and identify five key principles [7]:

1. Value
2. The Value Stream
3. Flow
4. Pull
5. Perfection

For the production and manufacturing environments, these principles added a significant degree of structure and effectively established rules for identifying value and removing waste (non-value add to the customer) from the process. In addition, with trust and respect for the individual as the heart of lean, the climate of the work environment made a natural progression toward a common organizational objective of "attract and retain the best." Most importantly, however, lean puts the individual at the center of the work. The elimination of waste through the principles of lean is only accomplished through the skills and capabilities of the people actually doing the work – the people who have been trained in lean methods so they can assess the process, find the waste, and eliminate it.

Lean Development – Increasing the Learning Rate

In Chapter 7, we introduced the concept of Lean Development which focuses on reducing waste that impacts a firm's learning rate. We can use Lean Development processes to increase individuals' control over their careers. To do this, we focus on how an individual does the work, is able to control their rate of learning, and the benefit a firm sees from this.

Exercise 10.1 Knowledge Acquired as a Function of Time – Shorter Development Time

To demonstrate the benefits of improving the learning rate, create a graph of Knowledge Acquired versus Time:

1. Draw a graph with X- and Y-axis, in which the X-axis (horizontal) is time and the Y-axis (vertical) is the Knowledge Acquired, with a range of 0% to 100%.
2. Draw a diagonal line at 30 degrees, beginning at (0,0), (y=0% and x=0) in which the endpoint of the line occurs when the y-coordinate is 100% – meaning, 100% of the knowledge required to deliver a design solution. The x-coordinate of the endpoint can be established as time=T_{30}.
3. Now, draw a second diagonal line at 45 degrees, beginning at (0,0) and ending when the y-coordinate is 100%, indicating that 100% of the knowledge has been acquired. The x-coordinate of this endpoint can be established as time=T_{45}.
4. What percentage of time=T_{30} is time=T_{45}?

Answer: You will find that our 45-degree diagonal line achieved 100% of the knowledge required in about 60% of the time compared to the 30-degree line (or about 40% sooner).

In the most general terms, the 30-degree line is traditional development and the 45-degree line is Lean Development. Or, in simple terms, Lean Development is better and faster. Here is the bonus though, through Lean Development, the 45-degree line, which achieved 100% of the knowledge required in 60% of the time, was achieved with fewer resources.

Research published in 1987 showed that Japanese automotive companies delivered, on average, a car in two-thirds the time with one-third the engineering hours [8]. From this, we can simplify the performance improvement to be: Lean Development is four times more efficient than traditional product development. For the remainder of this book, we will assume that any effort in the area of Lean Development is working toward a 4× improvement. This improvement is not the result of people working harder or finding people who are more dedicated, it is the result of the elimination of waste in the development system. To put it back into product development terms, it is people solving problems faster while delivering better customer solutions.

In Exercise 10.1, we focused on engineers and managers using Lean Development to create the "knowledge required" in ~60% of the time as compared to traditional development (the 45-degree line versus the 30-degree line). With this improvement, the business sees the direct benefit of an increased learning rate as it results in a product or service being delivered to a customer sooner. The benefit to the company is good, but now let's look at how that translates into gains for an individual.

Exercise 10.2 Knowledge Acquired as a Function of Time – Personal Growth

To demonstrate the opportunity for personal growth, establish the overall level of knowledge acquired by an individual over time.

1. The graph in Exercise 10.1, Time versus Knowledge Acquired, established the difference in learning rates between traditional (30-degree line) and Lean Development (45-degree line).
2. For the 30-degree line, we established that 100% of the knowledge required was established at time = T_{30} (on the *x*-axis).
3. Now, extend the 45-degree line until the *x*-coordinate of the endpoint is at T_{30}.
4. What is the increase in knowledge acquired on the 45-degree line compared to the 30-degree line? (i.e., what is the *y*-coordinate of the 45-degree line at T_{30} compared to the *y*-coordinate of the 30-degree line at T_{30})

Answer: In comparing the *y*-coordinate of the 45-degree line to the *y*-coordinate of the 30-degree line at time T_{30}, the 45-degree line (Lean Development) has 70% more Knowledge Acquired compared to the 30-degree line (traditional development).

In looking at Exercise 10.2, when we compare the 45-degree line (Lean Development) to the 30-degree line (traditional learning), we see an increase in acquired knowledge. The additional acquired knowledge drives an increase in personal development. In most cases, it is reasonable to assume that the individual working using Lean Development processes sees this overall increased learning because they finish one project in 60% of the time (from Exercise 10.1) and are then able to move on to the next project. In contrast, an individual doing the work in the traditional manner is still working toward time "T_{30}," in order to finish up the job. With benefits of Lean Development being seen by both the firm and the individual, we now need to establish tactical methods to implement it.

Based on our understanding that waste, in development, is anything that slows down learning and a root cause for low retention of women in engineering is lack of available knowledge, we propose that the tool to address both of those issues is the A3 Problem-Solving Process done via an A3 report.

ATTACKING BIAS AND BUILDING CONFIDENCE

Before we work through a method that increases both structured learning and structured problem-solving while simultaneously creating reusable knowledge, let's look at one different way of learning for a new-to-career engineer: learning by meetings.

TRADITIONAL: LEARNING BY MEETINGS

A new-to-career engineer sits down at her manager's desk to discuss her work on a new problem. That problem could be a technical issue that was discovered during product development work or it could be requiring a design for a new customer feature. The first step in the process is that both individuals open up their respective tools to capture notes of the meeting – maybe a notebook, laptop, or tablet. Being prepared for the meeting, they have both probably written down initial thoughts about the problem, questions, and maybe how to approach it. As they begin to talk through the problem, they are most likely capturing notes on their respective tools. Because neither can see specifically what the other is writing, they are totally dependent on the verbal interaction. However, the focused new-to-career engineer will naturally pay attention when the manager writes down something from the discussion, thinking that she should probably write down something as well. However, the manager could have written down something like "call mom" or "stop at store." As the meeting progresses, they establish a path forward and likely end the meeting by agreeing on and writing down (on their own tools) the specific next steps.

In some firms, the entire discussion may have been done via a whiteboard. The new-to-career engineer would then be able to see firsthand what the manager is writing down and would be able to take clear notes from the board. At the end of the meeting, the manager and engineer both agree on the next steps and can take a picture of the board. This approach can be a step forward, although it may feel more like a classroom setting to the new-to-career engineer than a coaching session.

In both of these examples, there is no form of reusable knowledge that they are creating together to help resolve the problem or move the problem forward for their next meeting. The next meeting will most likely occur in one week with e-mail exchanges sprinkled in until then.

For the next meeting, one week later, it starts the same as the first, the manager and the new-to-career engineer open up their note-capturing tool, start by reviewing the agreed-upon steps from last week, and repeat the process we went through above.

Let's be clear, we are proposing that this work method is on the 30-degree line of our graph that was created above (Exercise 10.1).

The A3 Problem-Solving Process

Now, let's work through the specific tool and problem-solving process of an A3. This tool is at the center of creating reusable knowledge, one of the root causes we found contributing to the low retention rate of women in engineering. Increasing access to reusable knowledge ultimately improves all three of the broad areas contributing to the low retention of women in engineering – improving work climate, job satisfaction, and an individual's control over their career. Addressing reusable knowledge through a specific process like an A3 has significant downstream effects.

The goal is to accelerate the learning process by making knowledge visible. The A3 Problem-Solving Process is a tool whose output is an A3 report – a visible representation of knowledge. A3 reports (or an 11 × 17-inch sheet of paper in the United States – A3 size paper in Europe and Asia) have been a central problem-solving tool for Toyota for decades. From Durward Sobek's and Art Smalley's book *Understanding A3 Thinking*, 2008, they write

> As we reflect on our experiences and research of Toyota, we find that intellectual development of people is a high priority at Toyota. We also find that Toyota uses the A3 reports system as a way to cultivate the intellectual development of its members, and the company management intentionally attempts to steer that development in specific ways.
>
> **[9, p. 12]**

Sobek and Smalley do an excellent job of explaining the A3 Problem-Solving Process and provide examples and methods for individuals to utilize this tool. The basic structure of an A3 report (on the single side of a single sheet of paper) captures clearly and concisely the situation, goals, potential solutions, implementation plans, and results. Ultimately, through the use of this sheet, which may occur over days, weeks, or months, the team arrives at a solution and the purpose of the sheet is then complete. In Chapter 11, we will further describe how this powerful problem-solving tool fits directly within the heart of Lean Development.

The purpose of an A3 report is to accelerate the solving of a problem and create reusable knowledge as part of this problem-solving process. Morgan and Liker's 2017 book, *Designing the Future*, has a clear and concise section, titled "Crisp, Clear Communication for Collaboration and Knowledge Transfer" that outlines the power of A3 reports. In addition, since writing their first book, in 2006, they write that, "we have still not seen a more effective tool for collaborative problem solving, communication, and learning if practiced with the A3 spirit of inquiry and continuous improvement" [10, p. 233].

The basic format of an A3 report uses the left side of the page to capture objectives, strategies, constraints, knowledge gaps, and so on, and the right side of the page to identify the specific learning events and the knowledge gained through them. (See Appendix A for the method of developing an A3 report and Figure A.2 as an example template.) The creation of this sheet facilitates dialogue focused on the flow of learning, aids team-based decisions, and involves a wider spectrum of the team.

As powerful as this tool is, some organizations may find A3 reports directly at odds with the normal use of slide deck presentations that are used during typical meetings. Addressing this conflict may be the most difficult area of change for widespread implementation of this tool within an organization.

AN A3 IS OUR CLOCK

Going back to our history lesson in Chapter 6 about the key to finding longitude being the clock, an A3 is our clock. We can picture each problem-solver (e.g., engineer) working to find a solution using an A3, just like every ship's navigator working to find the longitude using a clock.

In an effort to improve the training and learning process, let's look at how our previous story, Learning by Meeting, is approached with the use of A3.

LEAN: LEARNING BY A3 – A COACHING EVENT

With a new problem to solve, the early-to-career engineer sits down with her manager not just to talk about the problem but also to start solving it through the creation of an A3 report or, specifically, to use the A3 Problem-Solving Process. As the two begin their discussion, the identified problem content is written down on a piece of scratch paper laid out in the form of an A3. Through this discussion, the manager moves mentally into the role of a coach. By the end of that first meeting, the outline of the A3 report serves as the start of her A3. Additionally, before the engineer leaves the desk, the manager asks the engineer to send her a first draft of the A3 in the next day or so. The manager does this so she can review it and provide input to the sheet prior to meeting again. More than likely, the engineer will send the first draft by the end of the day.

For the next meeting, the engineer shows up with two printed copies of her partially complete A3. The process is clear, start with what you know, and work from there. As the meeting progresses, both are taking notes on their sheets. Although the engineer can clearly see what her manager is writing on her sheet, the engineer is not concerned with the exactness of the content because at the end of the meeting the manager gives the sheet to the engineer for her to use in updating the A3.

Just like in the story above, they agree on the next steps, the manager asks the engineer to send her an updated copy of the sheet within the next day or two. The manager will review the updated copy to ensure that the engineer is on the right track and moving the learning forward. Because the manager is directly aware of all the information on the sheet, the manager's review is at most a 5- to 10-minute activity, then a quick note back to the engineer with any additional thoughts or requests.

The next meeting begins like the second one, the engineer shows up with up-to-date copies of her sheet, and they quickly move the conversation from the previously agreed-upon steps to the learnings that resulted from those steps. Because they are both seeing the most current knowledge, the conversation time may take half the amount of time to align on what has been learned and what needs to be learned as compared to our previous method, "learn by meeting."

This process continues at whatever pace the problem requires and they meet as frequently as the new-to-career engineer needs in order to successfully accomplish the required tasks.

To be clear, we have now moved the learning rate to our 45-degree line (Exercise 10.1).

In our second story, between her meetings with her manager, the engineer is using her sheet to gather knowledge from her peers and learn. She brings this information back to her manager with whatever frequency the nature of the problem requires in order to drive her learning forward. Much like her meetings with her manager, the A3 is at the center of the discussions with her peers. The A3 is what others are using in order to provide additional coaching.

As we have previously described, when there is an emphasis on creating or making reusable knowledge, there is an increase in confidence and job satisfaction and mitigation of the negative consequences of bias. With ownership and control of the information, the new-to-career engineer experiences an increase in self-efficacy and confidence about their ability to do their job. The effects of bias are mitigated by the clear, written communication of knowledge and thought process on the A3, reducing the effects of unconscious beliefs about competence and ability.

ADDRESSING THE CHALLENGE OF RELATIONAL GENDER ROLES

With an understanding of the A3 Problem-Solving Process, let's return back to the challenge of managing relational work. When looking at the process of having a second engineer, typically a woman, complete the relational work in the development process (e.g., moving a completed design, done by someone else, through to implementation), it is clear that the inefficiency of "handoffs" has been built into the system. As described by Ward and Sobek, "A hand-off occurs whenever we separate knowledge, responsibility, action, and feedback" [11, p. 51].

Let's play out how a manager can coach an experienced male design engineer, who is accustomed to handing off his work to a female project engineer for implementation, using an A3 report.

AN ENGINEER'S RESPONSIBILITY: DETAIL DESIGN THROUGH IMPLEMENTATION – ADDRESSING RELATIONAL WORK

The manager may say to the male design engineer during a meeting, "It looks like your A3 does a good job of detailing out the objectives and purpose of your design work and how you are going to create it. The overall work to do that looks solid." The engineer responds, "Yeah, I wasn't sure how using an A3 would work out, but after I got partway through, it seemed to help me." The manager continues with, "Given your experience, I wouldn't expect anything else; however." Now for the experienced individual, "however" is a red flag word. For most people, very few good things in this type of discussion come after the word "however." This is like when negotiating a car sale and having the sales persons say, "good point, however" or having the word "however" come up during those parting words at the end of a first date.

The manager continues with the discussion, "however, your A3 doesn't cover the validation and implementation tasks. Do you have thoughts on that work?" The engineer gave some thought and then replied, "Well, that work is typically done by the Project Engineer and she does a very good job at it so I didn't include it." He then continues, "In fact, I am not even sure what she does or how hard it is. It just seems to happen." In an agreeing tone, the manager replies, "She certainly does a good job with that kind of work and it is not as visible as it should be; however, we are working to move the planning and execution of the work to the design engineers."

The engineer responds, "Yeah, that is what I have heard. So, I should send her my A3 and ask her to add in that work?" The manager clarifies, "No, we don't need to do that. She has worked with her peers and they have created a well thought out Process A3 for that work, which you can use for your sheet." The engineer replies, "so how do I do that?" The manager then explains,

> Well that step is simple, I will send you the link to the web location of their Process A3 and supporting tools so you can incorporate the work flow into your sheet. After you have a first pass done, we can review both your sheet and the new process as well as your new responsibilities and see if there are any holes and identify concerns. Does that sound good?

"I guess ok. That should work," the engineer replies.

Although we have written this story because it relates to the retention of women engineers, we need to be clear that this work can be viewed as a tactical step to address the waste within the development process due to the frequency of handoffs.

Progress Toward Improving the Climate

When we look at the changes in work climate in the story above, by addressing the work structure and, specifically, the training of the design engineer, we see two key changes. First, we have overtly moved what had become "gender"-driven work across the gender boundary. In this case, the boundary existed between the design work and implementation tasks. In addition, the design engineer will likely take a different perspective of the entire work required to deliver his design to the product level. Second, as the project engineer is now freed from managing the details of validation and implementation for every design effort, she has time available to pick up technical design content, effectively crossing back over the organization's self-created gender boundary toward design.

ANOTHER LOOK AT THE CAUSAL DIAGRAM

Based on our previous story, we can see the impact of the A3. Specifically, the male design engineers' A3 will eventually map out the entire flow of work. His work will likely be beneficial to a less senior engineer at a later date as he has created reusable knowledge. In fact, we might expect the woman engineer to use this in her newly added design work responsibilities.

We can also assume that we might have made some progress in the design engineer's possible bias about the value of relational work. For the female project engineer, when she picks up new design responsibilities, we may see increased confidence and an increase in personal and professional growth. She may even become a role model to other women. By our count, this straightforward change in the design of the engineering system could result in addressing ten of the influencers in our causal diagram.

Other Organizational Benefits

The work with an A3 also has broader, organizational implications. First, we have moved the design engineer to a broader level of system thinking. He is now in a position to look at his work all the way from customer need to delivery of his solution at the product level. This increases his contribution to the business. Second, by understanding the entire flow, he is now in a position to see how he might reduce waste within that flow of work. Third, with his sheet, he is now in a better position to coach less senior engineers in the work of design.

None of this would have occurred without the A3 unless the female project engineer was able to convince the design engineer to do his own work or the design engineer felt guilty about having her do the validation and implementation tasks of his work and volunteered to "help her out." Instead, a change occurred because the system was designed to drive the change. The manager's job becomes focused on executing the design of the engineering system.

CONCLUSION

In this chapter, we have explored the benefit of learning and implementing the structured use of the A3 Problem-Solving tool. This tool is used to both teach and change the process. We have also discussed the ways in which division of responsibilities along gender lines, i.e., technical versus relational roles, are related to our underlying gender schema and can reinforce bias, contribute to a negative work climate, and reduce job satisfaction. However, by changing the design of the engineering system with the implementation of Lean Development principles, like the A3 Problem-Solving tool, we can move the organization to a more gender-neutral division of responsibilities and increase access to knowledge resulting in increased job satisfaction and improvements in the work climate.

[Authors Note: An integrated graph of Exercises 10.1 and 10.2 can be found in Appendix A, Figure A.5].

REFERENCES

1. V. Valian, *Why so Slow? The Advancement of Women*. Cambridge, MA: MIT Press, 1998.
2. J. K. Fletcher, *Disappearing Acts: Gender, Power, and Relational Practice at Work*. Cambridge, MA: MIT Press, 1999.
3. C. Ashcraft, B. McLain, and E. Eger, "Women in Tech: The Facts," National Center for Women & Information Technology, 2016. Accessed: Dec. 03, 2021. [Online]. Available: https://wpassets.ncwit.org/wp-content/uploads/2021/05/13193304/ncwit_women-in-it_2016-full-report_final-web06012016.pdf.
4. P. F. Drucker, *The Practice of Management*. New York: Harper & Row, 1954.
5. J. A. Champy, and M. M. Hammer, *Reengineering the Corporation*. New York: HarperCollins, 1993.
6. R. Fry, B. Kennedy, and C. Funk, "STEM Jobs See Uneven Progress in Increasing Gender, Racial and Ethnic Diversity," Pew Research Center, Apr. 2021. [Online]. Available: https://www.pewresearch.org/science/wp-content/uploads/sites/16/2021/03/PS_2021.04.01_diversity-in-STEM_REPORT.pdf.
7. J. P. Womack, and D. T. Jones, *Lean Thinking: Banish Waste and Create Wealth in Your Corporation*. New York: Simon & Schuster, 1996.
8. K. B. Clark, W. B. Chew, J. M. Fujimoto, and F. M. Scherer, "Product Development in the World Auto Industry," *Brookings Papers on Economic Activity*, vol. 1987, no. 3, pp. 729–781, doi: https://doi.org/10.2307/2534453.
9. D. K. Sobek, and A. Smalley, *Understanding A3 Thinking: A Critical Component of Toyota's PDCA Management System*. Boca Raton, FL: CRC Press, 2008.
10. J. M. Morgan, and J. K. Liker, *Designing the Future: How Ford, Toyota, and Other World-Class Organizations Use Lean Product Development to Drive Innovation and Transform Their Business*. New York: McGraw-Hill, 2019.
11. A. C. Ward, and D. K. Sobek, *Lean Product and Process Development*, Second Edition. Cambridge, MA: Lean Enterprise Institute, 2014.

11 It Is Just Good Engineering
The Basics of Lean Development

There are a significant number of resources on the topic of Lean Development. This chapter's purpose is to provide the reader with an overview of Lean Development, a system-based change in the engineering process, with the ultimate goal of addressing the root causes, areas of opportunity, and major contributing factors identified in our causal diagram in Chapter 6 (Figure 6.2). Our use of Lean Development will draw from six principles. We will focus particularly on three of them: Reusable Knowledge, Set-Based Design, and System Designer. These three principles are core to changing how the work is done at the engineering and management levels. For more detailed information on Lean Development, see Appendix A – Going Home on Time, Lean Development – The Principles.

THE PRINCIPLES OF LEAN DEVELOPMENT AND THE BENEFIT TO BUSINESS

The shift within product development toward Lean Development is motivated by a desire to reduce inefficiencies: (1) how a firm gains the required knowledge to deliver its product or service, and (2) acquiring knowledge for future use. Lean, in general, is the elimination of "waste." Waste is anything that gets in the way of delivering "value-add" to the customer. At the most basic level, Lean Development centers on principles related to the growth of individuals/teams, the creation of value-add knowledge, and the flow of that knowledge or work. It is the knowledge about the solution that is moved through the supply stream and ultimately ends up as a product or service that a customer will pay for.

At the highest level of measuring waste, having only 15% of the engineering population being women rather than 50% is a waste of human capital. This waste is seen in the system as solutions take longer to develop and provide less customer value. In addition to the physical lack of human resources, companies also lack the diversity of knowledge and experience women engineers bring to solving problems. Dramatically increasing the knowledge available to all is a benefit to all. As we saw in our story in Chapter 5, in which the leader stated to the group, "It is the engineering system, used for decades that has created the problem and it is now the

DOI: 10.1201/9781003205814-18

engineering system that is going to address it." Lean Development gives us a foundation to change the engineering system.

Although a firm sees significant benefit in the form of improved delivery of a product or service to a customer, a woman engineer can see substantially more benefit from a change in how work is done via Lean Development. This is specifically through the creation of (reusable) knowledge and access to knowledge. We see this connection in our causal diagram (Figure 6.2), in which the lack of reusable knowledge is a root cause for at least four intermediary causes (role models, control of learning, technical depth, and confidence) and therefore is a contributor to all three of our identified broad areas contributing to retention (satisfying work, control over career, and work climate). With increased levels of reusable knowledge, we can now begin to model the work environment of a physician in residency. Knowledge is power.

Based on the industry's work in the area of Lean Development, we will use the following six principles to provide us a framework to change how the work is done and our three core focus areas.

The Principles*

1. The creation of reusable knowledge (*core focus area*)
2. Cadence, pull, and flow
3. Visual Management (added to Dr. Ward's original five)
4. Entrepreneurial System Designer (*core focus area*)
5. Set-Based Concurrent Engineering or Set-Based Design (*core focus area*)
6. "Build Teams of Responsible Experts" [1, p. 1]

As we covered in Chapter 10, we are working with the expectation that Lean Development, through these principles, moves us toward a 4× improvement in the performance of our engineering system.

As we review the principles, we have ordered them in such a way to build off of each other. To demonstrate this, in the simplest comparison, we know that you cannot build a team of responsible experts (Number 6) without those team members having access to knowledge (Number 1).

Additionally, we suggest that the first three principles can be viewed as providing infrastructure to the development process. These principles enable a leadership team to create a development system that supports the organization's broader objectives. The last three principles can be viewed as the principles that drive innovation and value creation, which means that they are what drives the development of a service or product. The System Designer must first understand both the customer and the business before a team of responsible experts can move from customer needs through to product delivery.

* Although in different words and a different order, five of the principles (not including Visual Management) were proposed by Dr. Allen C. Ward in his 2002 publication of *The Lean Development Skills Book*. © Dollar Bill Books, Ann Arbor, Michigan [1].

While all six principles are critical and form the building blocks of Lean Development, we will devote more time and space in this chapter to (1) Creating Reusable Knowledge, (4) Entrepreneurial System Designer, and (5) Set-Based Design. These three provide a foundation for an individual to take control over their career, whether the firm demonstrates leadership in empowering it or not.

1. Creating Reusable Knowledge – The Power of the A3 Reports

In Chapter 10, we identified how the use of the A3 Problem-Solving Process adds structure to the learning process and is our "clock" in solving the problem of a low retention rate of women in engineering. While our Chapter 10 story, Learning by A3, demonstrates its utility for a new-to-career engineer, the benefit of the A3 problem-solving process is independent of someone's experience. The tool forms a foundation for our first principle of Lean Development, Creating Reusable Knowledge. The method of an A3 is basically a system view of the problem, the work being done to find a solution, and the solution itself. We can think of it as a higher power version of a Free Body Diagram (FBD).* For those struggling with the use of an A3 report given 25 years of using slide decks in engineering, imagine, as an engineer, never having been taught how to construct a Free Body Diagram. Imagine never being taught to look at a system of force components and analyze their interactions. Imagine a third-grader asking you, "why does the block slide down the board in one way but the other way it tips over?" and replying with, "that is a good question, but I am not exactly sure why." Or, imagine that you are trying to explain to a teenager, who is complaining about moving furniture, that it is generally harder to push furniture across a carpeted floor than pull it but not having the thought process of an FBD to use to answer the question.

As we covered in Chapter 10, the power of an A3 report comes from the use of the single side of a sheet of paper that is structured in a problem-solving manner (e.g., the problem in the upper left corner) and the use of the sheet to capture plans, actions, and results in a clear and concise manner with clarity and succinctness. It is this sheet that the owner(s) of a problem uses to start the work they are responsible for. This tool is a powerful coaching tool. When an engineer reviews his or her sheet with a more experienced coworker, manager, or technical lead, there is an opportunity to coach that engineer in the thought process of her work and help guide the author to find the best solutions.

The use of an A3 report by a new-to-career engineer is the best way an organization has to replicate the learning process found in a typical university engineering class. It puts all the power in the hands of the author based on what he or she writes. It allows the author to effectively work toward a letter grade every time it is reviewed with someone and, most importantly, it allows the author to communicate her intent in a written form. The author then receives feedback that is specifically focused on growing knowledge in a methodical and structured manner, all via the sheet. An A3 can be reviewed against a holistic picture of the problem and the clarity of the

* A Free-Body-Diagram (FBD) is a graphical representation showing the relationship of the forces acting on an object.

thought process toward a solution. It can be considered done when it is worthy of receiving a letter grade of A.

Over the last nearly two decades, many resources have been created or published concerning the use of A3 reports; however, the message here is that when used well throughout an organization, they help to level the playing field for women. Although it is fair to assume that the primary purpose of an A3 report, as a problem-solving tool, is to capture and disseminate knowledge in a well-defined and controlled manner, the most basic purpose may be to bring the best out of people.

FINDING OPPORTUNITY

Using an A3 can help us find opportunities that may not have been visible without it. Taking advantage of these new opportunities can further career goals by enabling control over career.

As a basic approach to the creation of an A3 report, it is more than likely necessary to have a detailed spreadsheet to capture technical or business-level information. This spreadsheet can then be a convenient way to make graphs and charts that can be copied and pasted into the A3 report. These detailed spreadsheets become another form of reusable knowledge.

2. Cadence, Pull, and Flow

This principle is targeted at how the development work happens – how it starts, how it progresses through time, and what it delivers. This principle can be applied to the physical development of a piece of hardware that will be delivered to the customer or at the level of learning required to make a decision about a business or technical choice. At a product program level, this principle would be demonstrated on an A3 that maps out the overall development work and milestones that may occur over the course of many months. In addition, for efforts of smaller scope that a sole engineer may own, this principle enables someone to map out their work and identify knowledge gaps. The key to this principle, for the individual, is to think in terms of the following: what needs to be done and by when, what needs to be learned and by when, and what decisions need to be made and by when. To be clear, this is not done via a Gantt Chart. For more on that topic see Appendix A, Principle 2, and a more powerful approach Figure A.3 – An Integration Plan

3. Visual Management

A basic approach to Visual Management within a development process is to literally hang critical pieces of information concerning the development work on a wall at a central point within the development area. These documents hanging on the wall could be high-level program A3 reports, which help the development team communicate the overall work. The documents provide the knowledge needed by the team to do their work in a more cohesive manner. An approach created by Toyota for this principle is the use of an Obeya (big room) [2]. As the English translation indicates, an Obeya utilizes the confines of a room and its walls to provide a development team and its leaders a place to manage the work. A Visual Management approach can then create a management system within the development work, specifically one in which the management team uses the wall or Obeya to manage program activities.

While an Obeya may be helpful in changing the way the work is done at the engineering level, it may be a big step to do well at the beginning of a transformation. For our purposes, we will view the use of A3 reports shared with others, and specifically the management team, as a contribution to Visual Management.

4. Entrepreneurial System Designer

The person at the highest level of this principle has been typically called a chief engineer. The chief engineer is a thoroughly researched role within Toyota and is one that is now utilized across a variety of technical industries. At Toyota, this person fundamentally owns the delivery of the car. However, in our use of this principle to change the way the work is done, we are going to refer to this individual as a System Designer and frame it as a skill to be developed in every engineer and manager. This work method encourages the engineer or manager to look at the entire development work for which she or he is responsible for. In our story in Chapter 10, "Responsibility: Detail Design through Implementation," we broadened the designer's responsibility to include validation and implementation. We increased his ability to work at the system level by using the A3 he was creating. His manager did not just ask him to take a more systems-based approach to his development work and then send him off to do the work, he showed him how to do it through the A3 report the engineer was creating.

As we look at the efforts done by a System Designer, we can consider two elements of the work she/he does via our A3 process tool.

1. The A3 report represents a broad view of the objectives, the development work, and the individuals involved. This occurs by sharing them with others. Individuals can be directly involved with the work or adjacently involved. In these reviews, we are looking for not only alignment on the work but also areas of work that have been missed.
2. The second element is the "technical relevance" of both the A3 report being created and the nature of the discussions about the sheets. This can be assessed by asking the simple question, "Do the A3 reports guide the technical work?" If technical members view them as "management slideware," therefore not relevant to them, then that needs to be addressed. If the System Designer is doing her best to make them "technically relevant" and asking questions about the technical aspects, then those with the technical depth will naturally help move the process forward.

5. Set-Based Design – Change the Development Method

A second part of improving the learning process through Lean Development beyond the use of A3s is explicitly directed at how learning occurs. As covered in Chapter 9 and Figure 9.1, the classic development process is centered around Build–Test–Fix, in which the designers identify what the team believes is the best idea from the various options available and then "Build–Test–Fixes" the best idea to completion. Given enough time and money, the assumed best idea is most likely delivered to the customer. The use of the Build–Test–Fix cycle can create or amplify a negative climate and reduce confidence, toward the lone woman in the group. This can most easily occur within a team when the loudest voice is driving the decision toward what is believed to be the best idea. It is this type of situation that may negatively affect job satisfaction or a person's feelings relative to maximizing business contribution as the team struggles to deal with opinion over facts.

In contrast, the use of Lean Development focuses on "learning before design." Learning before design replaces Build–Test–Fix with Build–Measure–Learn [3]. This includes fully evaluating considered options before the design is selected while eliminating the weakest solutions which effectively moves the strongest design forward. This may also include maintaining a known solution even if that solution is not the most desirable. This known solution is what other solutions can be measured against as well as being an acceptable solution to provide the customer. Set-Based Design replaces Build–Test–Fix (Point-Based Design).

As shown in Figure 11.1, the use of Set-Based Design is a structured approach that is focused on learning. As mentioned in Chapter 9, the method of Set-Based Design was first understood from researching how Toyota managed product development [4]. In Figure 11.1, we see that the design team has identified six possible solutions, including identifying a known solution (solution Number 6). The purpose of each of the three learning cycles is to eliminate the weakest design, ultimately ending with learning cycle 3 in which the final (and strongest) design, Number 3, is selected as the chosen solution. All of this learning would have been done within the schedule and resources of the overall program. When the method is used as an integral part of the learning process, it takes less time, costs less, and takes less resources. It is as basic as: measure twice, cut once. Additionally, it has the ability to create a cohesive work group as the team works together through the learning process.

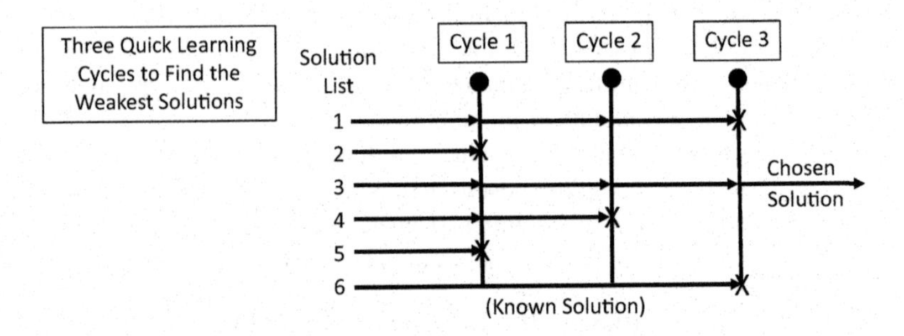

FIGURE 11.1 Set-Based Design – quick learning cycles.

IMPROVING WORK METHODS

As shown in our causal diagram (Figure 6.2), Set-Based Design helps us change the Work Methods resulting in increased Work Satisfaction. This structured approach, used within a team, helps to shift discussions from opinion-based to fact-based. It forces the team to focus on learning and, generally, learning together. It is a method that can mitigate bias and build a more inclusive team environment. Set-Based Design used at the individual level allows someone to establish their specific learning needs.

The business benefit of this learning method, and the resulting customer benefit, spans across a number of industries. However, this process has a unique benefit for women engineers working in an environment with various forms of bias. Specifically, the owner of a design area works with peers to identify design options using an A3. Rather than discussing which option is best, the discussions are focused on what learning needs to occur to eliminate the weakest design at any point in the learning process. Through each learning cycle, the designer can methodically identify the needed learning (sometimes referred to as knowledge gaps). Because this work is centrally focused on gaining the required knowledge to eliminate the weakest solution, the work is much less likely to depend on an engineer's personal insights and experiences. This is in contrast to Build–Test–Fix which is driven by personal insights and experiences and therefore can be heavily impacted by bias and increase confidence gaps. Although the use of Set-Based Design can be approached using defined a step-by-step approach, as shown in Figure 11.1, Set-Based Design is as much a thought process as it is a method.

WHAT THE INTERNET SHOWS US FOR SET-BASED DESIGN

If we were to conduct an internet search of Set-Based Design, we would find dozens of images that communicate the basic approach of starting with multiple options and then eliminating the weakest solutions through a learning process that leads to a final design choice.

This is an interesting comparison to our internet search of Science, Technology, Engineering, and Math (STEM) Engineering Design Process (Chapter 9). In that search, it returned dozens of colorful cartoon-type images of Build–Test–Fix designed for elementary-age children.

Trade-off Curves

Trade-off curves are a critical tool and thought process within Set-Based Design. Much like a causal diagram, which is a graphical representation of the causes of

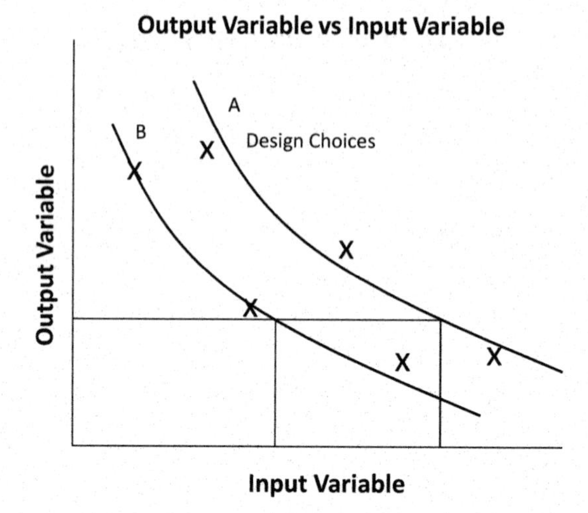

FIGURE 11.2 Trade-off curve.

a problem, a trade-off curve is a graphical representation of what the design space looks like.* They are intended to show the performance of a system. The most simplistic form of a trade-off curve uses an x–y graph with a performance variable on the x-axis, a customer-centric result on the y-axis, and a single line segment that runs through the graph. The line segment relates the performance parameter to the customer. Trade-off curves can be used to represent virtually anything that has some relationship between an input or situation and an output or result. The basic supply and demand curve, which relates the quantity (on the x-axis) to price (on the y-axis) is a well-known economic trade-off curve. Within product development, an organization would consider their trade-off curves as trade secrets. These curves dramatically improve the efficiency and effectiveness of the development process. Allen Ward wrote, "If I could teach you only one lean tool, trade-off curves would be it. Trade-off curves represent a basic way of thinking" [5, p. 153].

Figure 11.2 shows a basic format of a trade-off curve. An engineer's process of creating a trade-off curve as a tool in the development process naturally breaks the Build–Test–Fix mentality. It forces the discussion away from "what do you think will work?" to "what choices do I have and what do I need to do to identify them?" Our example diagram helps the engineer communicate the design space and choices to peers. This simple example, which may have been created by as few as six tests (three for each of the two design options), demonstrates how a trade-off curve can completely alter a discussion about which two designs, A or B, should be pursued. Like the general purpose of an A3, it moves the discussion from opinion and bias to facts. In our example, design A and design B may fundamentally differ in cost and/or technical risk. The graph helps the discussion start with "what do we need to do

* Allen Ward and Durward Sobek demonstrate the linkage between a causal diagram and trade-off curves in their book *Lean Product and Process Development* (2014), 2nd Edition [5, p. 159].

to move the Input Variable to the left in order to allow us to move from design A to design B while maintaining the same Output Result." Trade-off curves are another tool grounded in good design practices that deliver more customer value.

INCREASING TECHNICAL DEPTH AND CONFIDENCE

In our causal diagram (Figure 6.2), Trade-off curves are a tool to drive confidence and technical depth resulting in increased work satisfaction and control over career.

Figure 9.2 is an example of a trade-off curve. It shows how work structure drives work climate relative to how progressive a firm is. A firm can use this type of curve to establish where to invest its efforts and what kind of result to expect.

Let's look at a hypothetical example of how to coach an engineer in the use of Set-Based Design, how to teach someone to break the cycle of Build–Test–Fix, and how to learn about the design space through trade-off curves.

COACHING SET-BASED DESIGN AND THE USE OF TRADE-OFF CURVES

A relatively new-to-career engineer is asked to resolve a technical design problem that the program team recently discovered. When the engineer sits down at her manager's desk to talk through the work, she is expecting to start discussing which design change seems to be the best. However, the manager is changing the way the work is done. She starts the discussion by asking what options the engineer thinks exist and what needs to be learned for each of them. Toward the end of the discussion, she asks the engineer to start an A3 with the basic elements: objectives, the situation, the required implementation schedule, and a learning plan for the possible solutions. To help with that request, the manager sketches out the basic format of the A3 on scratch paper. The engineer leaves the meeting with the sketch of the A3 in hand.

The engineer starts her A3. She begins learning from her peers about the current situation, the objectives, and the needed timing for the solution. She learns that this issue was found in the program's customer-centric system-level test, but there is little detailed knowledge beyond that and little knowledge about the design is written down. During those discussions, the engineer may get input from her experienced peers that the solution is pretty straightforward; however, she understands that she has been asked to do the work differently and proceeds through the steps of establishing the learning needed for the solution space, building her A3 in the process.

When the engineer brings her A3 to the manager, they start the detailed discussion of reviewing the issue and current situation. They then proceed to discuss the options and the knowledge needed for them – marking up the A3 as they talk. Some design options can be combined. Through this discussion, another option is identified. Eventually, the discussion moves to a very specific change in expectations for the engineer.

Typically, for the organization, the learning process of how a design performs is done through the customer-centric system-level test that found the issue. But in this customer-centric system test, the manager considers the finding of a product failure to be a failure of the development process. However, to set this expectation, without teaching someone how to approach the task is wrong. So, the manager moves the discussion toward the use of trade-off curves to understand the overall design space and explains how to approach them.

For each learning area, they talk about the axes of the trade-off curves, what the performance curve inside it might look like, and how to approach creating the curve. They agree that if the engineer establishes the two endpoints and the middle for any line on the graph, she will have enough information to compare the various options. They end the discussion by agreeing that the next step will be to create the trade-off curves and relate them to the system performance requirements.

Through the work of creating the trade-off curves, the engineer is conducting just enough learning to find and exclude the weakest designs, sometimes with a single test, and with each learning she moves the strongest solution forward. She eventually arrives at a solution that is proven to meet the requirements with an implementation plan that meets the program objectives. The engineer reviews the solution with her manager and others. Through those reviews, the design is finalized and incorporated into the system-level test and the engineer moves on to the next opportunity.

6. Teams of Responsible Experts

In Allen Ward's and Durward Sobek's book *Lean Product and Process Development*, they list the four tasks of a responsible expert: "Focus on overall project success, create new knowledge, communicate it, and represent it (conflict to consensus)" [5, p. 227].

As the leaders in the organization move individuals toward the use of A3s and Set-Based Design, they naturally grow each individual's capability. As individuals begin to share their knowledge with others through the A3s they are using, they naturally increase each other's knowledge. Finally, as the entire organization moves down the Lean Development path, teams of responsible experts are created. We place this principle at the end of our list because this is the result of the effort of the preceding five principles. In Chapter 10, we began our discussion of Lean Development from the perspective of increasing the rate of learning of those doing the work. To put that in context, we used Exercise 10.1 to show the business benefit of someone being on the 45-degree line and then demonstrated in Exercise 10.2 the personal gain that an individual sees. It is this personal gain at the individual level that contributes to creating teams of responsible experts.

If we reflect back to our Soccer Parent story in Chapter 9, the six principles establish the rules of the game and what to expect of the players or, in our case, the engineers and managers delivering the engineering customer-valued solutions.

HOW LEAN DEVELOPMENT FITS INTO THE DEVELOPMENT PROCESS

The basic product development approach, in which a firm works to understand customer needs and to identify solutions to fill that need in a manner that is profitable for the firm, is still maintained as the key basic steps. However, how that work is actually done is where there is significant difference between traditional product development and Lean Development as "Lean development's goal is learning fast how to make good products" [5, p. 14]. This goal fundamentally alters the approach to product development. Lean Development identifies specific knowledge gaps and establishes concrete methods to fill those gaps.

The beginning of the work might be to fully understand customer needs or a gap that you are trying to fill – a need that may be unknown to the customer at that time. The A3 tool is a concrete method to accelerate that initial learning to clarify customer needs and customer benefits and translate that into a profitable value stream. As Jeffery K. Liker writes in his book titled *The Toyota Way*, Principle 13, "Make decisions slowly by consensus, thoroughly considering all options, implement rapidly" [2, p. 237].

Coming back to our critical tool of A3s, briefly, we will describe four basic uses of A3 reports.

A3 Reports – Problem-Solving, Opportunity, Planning

These sheets may be constructed to specifically solve a problem that the organization needs to address, an opportunity that might exist, or a plan to be created. The left side of the sheet communicates the current situation, goal, and so on, with data and graphs that are currently known or understood. The right side of the sheet identifies possible solutions or corrective actions plus learnings. The sheet is used through the learning process to help the team arrive at a solution or plan that meets the goal.

A3 Reports – Integration Plans: Manage the Flow of Work and Learning

These sheets specifically map out the learning and workflow, focusing on Build–Measure–Learn. They reinforce the idea of Measure-Before-Design and Set-Based Design. These sheets can be used at any level within the organization and at any level of product development. In addition, the work consists of quick learning cycles that eventually all flow into a customer representative design that is integrated on the "very right side of the page." These learning or integration events may cover many months of work, so the A3 is a working document. The highest level of learning, at the customer level, naturally cascades to more detailed learning sheets.

A3 Reports – Status Sheets

These sheets may be used to summarize the current situation of the development work or customer-centric data that has been collected. When used in this way, they are typically either input to other A3 sheets or output from them. However, they may be a primary A3 used within decision processes.

A3 Reports – Knowledge Sheets

The A3 knowledge sheets are created to support the creation of needed learning. In most cases, they are driven by specific areas of design or customer learning and focused on gaining the required knowledge. These sheets are used to structure the discussions with others and capture the knowledge. These sheets enable the coach to ask the author, "what other solutions have you considered or what other knowledge do you need to maximize your solution to the customer?" Most importantly, they give the author a written voice that may be free of bias and create confidence. For the coach, this written word can minimize implicit bias when it is constructed with an appropriate level of confidence and with the best knowledge available. Ultimately, they enable the design work to deliver the highest possible business benefit.

The A3 reports typically sequence from broad to narrow in scope. They typically begin with opportunity sheets, then transition to learning or integration sheets, and finally to an A3 knowledge sheet. Each of these sheets creates reusable knowledge for future development work.

CONCLUSION

There are many resources describing the use of A3s, working as a System Designer, the use of Set-Based Design, and the use of trade-off curves. When they are used well throughout an organization, they "level the playing field" and deliver better business results than without them. Most importantly, they can be used at the individual level, increasing an individual's opportunity for personal growth, increased technical depth, and ultimately increased control over their career and increased job satisfaction. All of these factors can increase the retention of women in engineering.

REFERENCES

1. A. C. Ward, *The Lean Development Skills Book.* Ann Arbor, MI: Ward Synthesis, Inc., 2002.
2. J. K. Liker, *The Toyota Way: 14 Management Principles From the World's Greatest Manufacturer.* New York: McGraw-Hill, 2004.

3. E. Ries, *The Lean Startup: How Today's Entrepreneurs use Continuous Innovation to Create Radically Successful Businesses*, First Edition. New York: Crown Business, 2011.
4. D. K., Sobek II, A. C. Ward, and J. K. Liker, "Toyota's Principles of Set-Based Concurrent Engineering," *Sloan Management Review*, vol. 40, no. 2, pp. 67–83, Winter 1999.
5. A. C. Ward, and D. K. Sobek, *Lean Product and Process Development*, Second Edition. Cambridge, MA: Lean Enterprise Institute, 2014.

Introduction to Chapter 12

Alissa R. Stavig and Robert N. Stavig

In the first 11 chapters of this book, we described three major factors contributing to women leaving engineering, specifically unsatisfying work, lack of control over their career, and a negative work climate. We further identified three specific root causes: a lack of reusable knowledge, the way the work is done, and a lack of role models. Up until now, we have focused on implementing lean processes to change the mechanics of development work to increase the amount of reusable knowledge and change the way the work is done. We now shift our focus to our third root cause, role models, and explore the ways in which professional relationships are critical for increasing work satisfaction, improving the overall work climate, and increasing an individual's control over their career. For leaders chartered with increasing the retention of women in engineering, this chapter should provide an understanding of key professional roles that facilitate both individual and organizational development and growth.

Given the nature of this chapter, we have asked two skilled women engineers with a broad breadth of experience, Deb and Rose, to provide their insights. The information in this chapter is based on research and recommendations they have gathered from others with examples from their own personal experiences as well as interviews they conducted with the consent of women in their professional network. Quotes and anecdotes are drawn from interviews and illustrate general themes from individual women's careers. Names of interviewees have been changed when first names only are used.

Of note, this chapter focuses on the ways in which technical coaches and mentors promote personal and professional growth in the workplace through mutually beneficial relationships based on trust and accountability. Mentors and coaches can serve valuable roles in our personal and professional lives including providing emotional support along the way. However, coaches and mentors should not replace professional mental health treatment if it is needed. Specific reasons that an individual might need to seek mental health treatment include, but are not limited to, having difficulty functioning at home or at work; having difficulties in relationships; having changes in behavior, emotions, or thinking; and having suicidal thoughts [1, 2]. If you are in an emotional crisis or having suicidal thoughts, you can call the National Suicide Prevention hotline, 1-800-273-8255 [3]. If you are seeking professional mental health treatment, you can call SAMHSA's National Helpline, 1-800-662-4357 [4].

DOI: 10.1201/9781003205814-19

REFERENCES

1. American Psychiatric Association and American Psychiatric Association, Eds., *Diagnostic and Statistical Manual of Mental Disorders: DSM-5*, Fifth Edition. Washington, DC: American Psychiatric Association, 2013.
2. "What is Mental Illness?," American Psychiatric Association, Aug. 2018. https://www.psychiatry.org/patients-families/what-is-mental-illness (accessed Feb. 12, 2022).
3. "National Suicide Prevention Lifeline," National Suicide Prevention Lifeline. https://suicidepreventionlifeline.org/ (accessed Feb. 12, 2022).
4. "SAMHSA's National Helpline," SAMHSA (Substance Abuse and Mental Health Services Administration), Feb. 08, 2022. https://www.samhsa.gov/find-help/national-helpline (accessed Feb. 12, 2022).

12 Accepting a Hand Up

Role Models, Technical Coaches, and Mentors

Debra Blakewood and Rose Miranda Elley

Take a moment to reflect on the personal inventory of professional relationships in your life. If you have not yet established a role model, technical coach, and mentor, now is the time to start. This chapter is committed to explicitly defining and deeply exploring these relationships, and will provide helpful insights to deepen your engagement.

As you read this chapter, we recommend adopting the perspective of developing these relationships not only to support you in your career but also from the perspective of ways you can support others. Rose's mother taught her in life there will always be people who are ahead of you and there will also be people who are behind you; however, life is not a zero-sum game with winners and losers. Supporting others can be mutually beneficial – facilitating connections with others, enhancing our engagement and satisfaction at work, and increasing our sense of belonging. Seeing where you are in the grand scheme of things is empowering, serving as a reminder of how far you have come and suggesting how much farther you can go.

The field of organizational science introduced the concept of psychological safety, "the belief that the workplace is safe for interpersonal risk taking" [1, p. 114], as a factor contributing to individual growth and development in the workplace. Having a culture of group trust, composed of both cognitive trust and affective trust, has been described as critical for psychological safety [2]. Cognitive trust is established when there is belief in an individual's capability and consistency, capability being the knowledge and skills to get work done and consistency being dependable performance in a reliable manner [2]. Affective trust is the belief in an individual's integrity and their ability to be caring, empathetic, and warm [2]. Mentors and technical coaches combine cognitive and affective trust and are an important foundation of psychological safety in the workplace, thereby enabling an individual's maximal performance and satisfaction at work.

Role models, mentors, and technical coaches all have unique roles in our professional development. Our chosen role models are our North Star, guiding us; our mentors help propel us toward our goals. We select our role models, whom we may or may not meet, based on life, work, and virtues we are inspired to emulate. In contrast, we create relationships with mentors who guide us to greater success with direct communication and knowledge sharing. These pivotal relationships can be

DOI: 10.1201/9781003205814-20

transformative in building our self-confidence, establishing our habits, growing our network, and expanding our personal vision of what we will accomplish.

DEFINITIONS

Technical Coach: An individual with either similar or greater technical knowledge with whom you can comfortably discuss technical issues and knowledge gaps; the nature of the relationship tends to be instructional [3, 4].
Role Model: An individual you aspire to emulate [5].
Mentor: An individual who works with you to understand your goals, helping to guide and support you along your career path [4, 6].
Sponsor: An individual who recognizes your technical prowess and has the level of authority to open doors for you through key introductions, political connections, and assignment of resources (financial or personnel) specifically for your advancement [6].

Ideally, for our roles as technical women, we have found role models and mentors throughout the course of growing up, during college, and in our early-career years. Mentorship often happens organically with families who have historically geared toward math and science, friendships with peers who share our technical passions, or professors who teach us to solve complex problems. In addition to role models and mentors, having relationships that fulfill the role of technical coaches and sponsors are key to career success [6], with several successful women in Science, Technology, Engineering, and Math (STEM) commenting on their importance in professional identity formation and career success. Notably, mentors and technical coaches can both take on the role of sponsor.

Foundational relationships with role models, mentors, and technical coaches create the groundwork necessary to achieve success and satisfaction. Abraham Maslow, a world-renowned psychologist who studied exceptional people as well as other groups of people, described five hierarchical needs driving motivation and performance: physiological needs, safety, belonging, esteem, and self-actualization [7], [8]. While criticisms have been made of the model in the 80 years since it was published [8], in creating a positive culture, creating an environment of safety and belonging remains of utmost importance. Strong mentors and technical coaches clearly facilitate belonging and can contribute to esteem.

The specific ways in which a positive role model or mentor can contribute to psychological safety are further demonstrated by the stages of psychological safety described in Tim Clark's book *The 4 Stages of Psychological Safety: Defining the Path to Inclusion and Innovation* [9]. Clark, a social scientist, defines Stage 1, Inclusion Safety, as "Respect for the individual's humanity" [9, p. 103]. Stage 2, Learner Safety, is "Respect for the Individual's innate need to learn and grow" [9, p. 103]. Stage 3, Contributor Safety, is "Respect for the individual's ability to create value" [9, p. 103]. Lastly, Stage 4, Challenger Safety, is "Respect for the

Individual's ability to innovate" [9, p. 103]. Stage 4 is where innovation occurs and where companies "redeem" the value of their diversity. By building trust within professional relationships, supporting an individual's ability to learn, and providing opportunities for individuals to give back to others, technical coaches and mentors can create an environment of psychological safety fostering personal and organizational growth and development.

ROLE MODELS

The textbook definition of a role model is someone whom we admire and aspire to emulate [5]. As we go through life, there are many people whom we may strictly admire. These are not role models because they do not impact our behavior or our life choices; we do not see ourselves in their image or desire to follow in their footsteps. However, when we emulate someone, we take our admiration to the next level. We identify who we would like to become. We begin an internal transformation process through a now-established standard we have selected to guide our decisions. In addition, at conscious and subconscious levels, through our intention to emulate, we have acknowledged to ourselves that we have both the capacity for achievement along the same lines as our role model and a propensity for self-awareness such that we can incorporate traits into our being [10]. When discussing her personal definition of a role model, mining engineer Audrey told us, "a role model is someone who embodies the professional and personality traits of a successful individual where you can say, 'I like how they did that and I want to embody that.'" This belief in our capacity to emulate fuels healthy self-confidence and self-efficacy.

As women in engineering, our choice of role models can be a window into the level of confidence we have about our potential for advancement either technically or on the managerial track. Have we identified role models to provide direction, guiding us to become the best version of ourselves, or have we shied away in fear?

Several women we spoke to shared fond stories of family members in engineering and science who were their first role models. Paula, an electrical engineer who moved into management after only five years in engineering practice, shared that her father had been a Vice President of manufacturing. She stated, "Because he did it, I knew I could do it, too." Her father was a lifelong role model. Ashley, an earth scientist, shared that her mother had been a computer scientist who had encouraged Ashley to become an engineer. As her role model and also a mentor, Ashley's mother serves in both a professional and a personal relationship. Veronica grew up in a "science-y" family. Her father, aunt, and uncle were engineers, and her mother was a programmer. The decision to become a mechanical engineer was a tough one for Veronica, but her early-established passion for the sciences, fueled by her knowledge of her role models' successes, prevailed.

Peers can also be early role models. In school, Lorin was majoring in psychology but she had an aptitude for math. During a study group, her male peers shared that they were pursuing engineering. Interest piqued, Lorin transferred majors into engineering, soon finding a co-op summer job working for an aerospace contractor.

Madison, a space environment specialist, adjusted her undergraduate degree path after attending a conference and meeting a peer engaged in interesting work. She modeled her ongoing education and training to follow the same path. Brooke, a highly accomplished engineer, had high school physics and calculus teachers who recognized her ability and suggested she study engineering in college. Driven by a childhood passion for watching space launches, she earned scholarships and graduated with her degree in aerospace.

The importance of early exposure to trusted familial and peer role models highlights the special challenges facing women who are the first in their family to pursue entrance into engineering, and likely contributes to disparities in recruiting those from racially and economically diverse backgrounds. The critical nature of exposure to career options was demonstrated in a study of inventors. By examining the childhood backgrounds of patent holders, the study found that exposure to innovation during childhood through parental or neighborhood influences had a significant impact on the likelihood that a child became an inventor [11]. The study's authors, Bell et al., concluded that low exposure to innovation contributes to a low diversity of inventors in terms of gender, socioeconomic background, and race/ethnicity [11]. This study demonstrates the importance of role models in influencing career decisions and therefore the barriers to overcome when working to increase diversity in engineering.

Despite documented benefits of aspirational role models, many women in engineering are not continuing to identify role models as their career progresses. We believe that role models are important for women at all phases of their careers.

Rose has found through her own professional journey that even the most successful individuals go through periods of struggle and self-doubt. These times call for deep self-reflection. Who are you? Who do you want to be? Through your self-exploration, identify your goals and values. What is important to you? What is your style? Assess the gaps between where you are now and where the role models are that you seek to emulate. Then, seek to understand these gaps in the context of your self-confidence. In the article, "The Relation Between Chosen Role Models and the Self-Esteem of Men and Women," Wohlford et al. state, "Role models may 'provoke self-enhancement and inspiration when their success seems attainable, but self-deflation when it seems unattainable.'" [12, p. 576]. In other words, it can be helpful to understand if avoidance is occurring. The more personal work you invest in exploring the roots of your self-confidence and your ego, the easier it will be for you to identify what will help you build confidence as you work toward your goals. Keep yourself open to seeing the role model standing in front of you, and be ready for growth and learning opportunities. [13].

The culture of the professional environment has an effect on your confidence. Consider these well-described workplace challenges: gender bias, imposter syndrome, and diminished psychological safety [1, 14, 15]. Next, consider how your work teams actively practice psychological safety [9]. At what stage is your organization operating? Is trust, both cognitive and affective, present to foster your professional development? Once you understand the significant external forces impacting your confidence, it may be instructive to revisit your self-assessment.

You can find a role model almost anywhere you look [16]. You do not need to choose the most successful individual as a role model. Everyone is inspired differently. One R&D engineer, Cathy, watches the people around her who are doing good work and incorporates or emulates what she finds to be their aspirational best practices.

As female engineers, we all have the potential to be role models. We have a unique and privileged opportunity to support younger women in engineering. Others we have spoken to feel similarly. Sabita, as a hardware engineer, noted, "I am a role model for other women." Principal Engineer Johanna told us, "I feel a lot of ownership about how I can make this easier for the women behind me. How do I clear the path and make sure as they come up they do not go through the challenges that I went through?"

Serving as a role model can facilitate authentic and satisfying achievement. Recognizing her important role as a model for others, Sabita aims to model servant leadership with inclusivity and authenticity. Together with a few colleagues, she founded a network whose actions led to a more gender-inclusive work environment. With the grassroots creation of this safe space, her actions catalyzed the entire organization to reach a higher level of psychological safety. Though each woman in engineering will have her own special niche and unique interests, a common theme is the importance of awareness and intentionality.

TECHNICAL COACHES

Technical coaching focuses on the technical aspects of work, developing problem-solving, expertise, and excellence as compared to the other types of professional relationships [3]. Technical coaches feed technical depth and increase confidence (Figure 6.2). Everyone needs technical coaching when they are starting out [3]. Many firms offer such specialized technical training. If the firm lacks a formal program, a mentor or manager should facilitate and foster the connection. The manager's training should include how to identify knowledge gaps, make them visible, and, most importantly, ensure the result is without negative repercussions [3].

Annette, an electrical engineer and founder of three companies, reflected back to us on the start of her career. "Absolutely … it would have been helpful to have a technical coach – where you feel safe, and they make it safe." She recalled that when she had reached out to technical leads, "It didn't feel safe to ask them questions. I wanted to ask them to begin at step zero when they are talking at step 150." At the same time, individuals may feel pressure or receive messages to stop asking questions, such as when Deb received a stinging, "I've helped you enough." Neither Deb's psychological safety nor her performance was likely enhanced.

Having a technical coach to provide specific recommendations based on their professional knowledge and experience can be particularly important for individuals from groups who have been marginalized. For example, five years after earning her master's in mechanical engineering, Johanna passed the Professional Engineering (PE) exam at the urging of her mentor, also a certified PE. While a PE is unnecessary for most jobs, her mentor observed that women and individuals from marginalized groups received more pushback and questions about their technical depth during technical promotion discussions. In contrast, certified women and men would sail through those discussions. Johanna is now a Principal Engineer at a Fortune Top 100 company and believes her PE certification helped open doors. She now encourages her protégées* to follow her example.

Firms that formally embed well-done technical coaching benefit from improved employee retention and productivity as well as increased efficiency and engagement [17]. Supported engineers will more quickly build confidence for facing technical challenges and will be more satisfied in their jobs. A well-designed program allows both parties some choice in the pairing to ensure good chemistry [17]. We have described other features of successful formal coaching or mentoring programs, based on a Catalyst report by Donolfo and Nugent, in the text box [4]. The success of the relationship hinges on the protégée's feelings of safety, but again, there needs to be a foundation of mutual trust.

FEATURES OF A SUCCESSFUL FORMAL COACHING OR MENTORING PROGRAM [4]

- Pair individuals based on their training and development needs
- Set clear expectations regarding time commitment
- Clearly delineate goals and monitor progress toward goals
- Create accountability

Specific to coaching, we highlight a few of *The Important Site's "10 Reasons Why Trust Is Important"* [18]. Trust facilitates communication and honest discussions, increases self-efficacy and positivity, decreases stress, and allows for meaningful connections which give the protégée a sense of belonging [18].

Ways to build mutual trust [19]:

- Maintain confidentiality
- Honor boundaries
- Be honest
- Be authentic
- Be responsible; admit errors, and correct them
- Follow up

It is clear that while protégées receive educational and experiential benefits, technical coaches also stand to benefit from the relationship by improving their own communication skills and ability to motivate employees, increasing their interpersonal and professional networks, and giving back to the profession [3].

* We use protégée, the female version of protégé, due to its international use and reference to women – in preference to mentee, which is specific to the United States and focuses on the mentor/mentee dyad.

Tips for a Good Technical Coaching Relationship

Susan Solomon, PhD, coach for Senior Tech Executives, told us that a mentor or technical coach should "truly care about their people and want to develop them." Ohio University's blog on technical mentoring provides additional recommendations for other best practices for technical coaches [3]:

- Lead "by example in problem-solving, prioritizing, and decision-making" [3]
- Explain "the thought process they use when resolving technical issues" [3]
- Discuss "technical tools and methodologies" [3]
- Review technical plans and detail designs [3] (which we propose is done via A3s and incorporating the principles of Lean Development into the work)
- Help their protégées to "identify technical goals and knowledge gaps" [3]

Linda Phillips-Jones, PhD, in her booklet *Skills for Successful Mentoring* [19], developed a framework of mentoring skills that applies to the technical coach/protégée dyad just as it does to the non-technical mentor/mentee relationship. She emphasizes the importance of bidirectional good listening. This can look like being a sounding board for ideas, listening deeply and attentively, and practicing active listening. A good listener also signals they have processed what they heard [19]. Phillips-Jones describes what follows: "as a result, they feel *accepted* by you, and trust builds" [19]. We encourage *authentic* listening by also avoiding interruptions, paraphrasing as needed, and, most important, listening with a receptive and open mind [19]. While calmly responding, an authentic listener can encourage out-of-the-box thinking. Avoid judging and replace "that won't work" with "how will that work?" At the same time, protégées can effectively communicate by specifying whether they understand the information they are being taught, what their goals are, and what help they need [19].

Giving encouragement is an important component of the coaches–protégées relationship [19]. "It's important, the congratulations," says Zaynab, one woman we spoke to. Jill S. Tietjen, PE, who spent her career in the electric utility industry and now sits on corporate boards, credits her achievements to "a lot of support and encouragement as well as a TON of hard work on my part."

The guidance a technical coach gives should be tailored to the protégée based on their goals and who they are as a person – both at work and outside of work. Tailoring your approach as a technical coach not only improves the chance that your protégée will be able to successfully implement your guidance, but also demonstrates empathy and caring about your protégée as a person. For example, Lorin, an industrial engineer, invested more time mentoring Kaya, her introverted industrial engineering protégée. Kaya perceived Lorin's empathy and commitment, and the payoff was tremendous growth.

Intentional goals shaped by reality are important for both sides in the coach–protégée relationship [19]. We are all on a journey. Some of us have just let it happen, e.g., one retired engineer described her career path as that of a frog trying to cross a river, jumping from lily pad to lily pad. A contrasting approach is to develop an intentional navigation system. Several engineers we spoke to were intentional about their goals and discussed them regularly. They were excited about their careers and proud that their achievements had far exceeded their "starter" expectations.

Technical coaches also need to provide prompt, constructive, and specific feedback that will help move the protégée to a more positive outcome [19]. For feedback to be effective, it must be delivered in a sensitive and thoughtful manner: illuminating the positive and the negative, delivered in a private setting when the protégée is ready to receive feedback, and using a measured tone [19]. Research suggests it is better to provide a higher amount of encouragement than corrective feedback [19]. Like a concerned parent, the technical coach needs to decide whether it's best to brainstorm better ways or just tell their protégée how to better handle situations. It is never fun to receive negative feedback; however, a technical coach, someone who is invested, will likely be gentler than other people [19]. One way a technical coach can improve the coaching relationship is to solicit feedback by asking, "Are you meeting your goals?" [3].

Another important role of a coach is to provide both specific expertise and broader insight to allow the protégée to gain a deeper understanding of issues [3]. Knowledge is precious, but does not decrease in value when shared; it is not meant to be hoarded or gated. Support and investment in the protégée's growth across all career stages are important. Sometimes people may need more specific direction and guidance. At other times, they need their mentor to demonstrate their trust in them by taking a step back, avoiding micromanaging, and allowing increased responsibility [3]. In some situations, protégées may need primarily support and validation; while in other situations, they may need to be challenged.

Protégée Responsibility

For the protégée, there are several behaviors that enable and engender success: action, quick learning, and follow through [19]. A successful protégée is an active one. The protégée should put herself in spaces where people are looking for emerging talent and should exhibit characteristics that will benefit the profession. A strong motivation to succeed is attractive to mentors and technical coaches. Dependability is an essential work trait. When working with their coach, protégées should demonstrate follow-through by documenting and keeping agreements [19].

Taking responsibility for directing your own educational and career development helps others help you [20]. Once you can articulate your tentative goals, your strengths, your development needs, and the assistance you would like, it is easier to both choose and influence your team of technical coaches, mentors, life coaches, sponsors, and advocates [19]. The people at the top of their fields have made themselves visible, demonstrating their value to be worth the investment. Ultimately, the protégée should be the one who takes initiative to make sure the help they receive aligns with their goals [19]. Another responsibility protégées have is to initiate discussions explicitly determining the preferred method of communication between meetings and how frequently meetings will occur so that both the mentor and the protégée feel comfortable in the relationship in whatever form that takes [3].

It is possible to over- and underestimate our performance as well as how others perceive us and relate to us. Honest reflection is the final protégée responsibility and to practice it, it can be helpful to try viewing yourself from your coach or mentor's perspective. Where are your areas of discomfort? Perhaps people may not see you as you want to be seen. Brace yourself because it is possible not everything you observe will be flattering. Rather than seeing negatives, your job is to consider weaknesses as an area for growth, which you can focus on within the context of the technical coach–protégée relationship.

If you cannot find a technical coach in your firm, help exists outside of the workplace. Connections available through professional and technical organizations offer other paths to learn skills. New businesses are popping up and their coaching expertise can be hired by the firm or engineer.

The idea of embedding formalized technical coaching may be met with resistance, but the short-term benefit of psychological safety and self-awareness leads to long-term benefits in terms of job growth and satisfaction.

MENTORING

Firms seeking outstanding performers should support mentoring of all types: technical coaches, values mentors, life coaches, sponsors, peer mentors, and employee networks. It is widely believed that mentoring works for everyone, but generally male colleagues have historically had more and greater access to mentors [4]. Those lacking mentors may experience a longer learning curve, receive less career advice, and miss the feeling of truly belonging [16]. However, it is important to find the right mentor. Some firms have established "formal" mentoring programs; officially sanctioned, recognized, and supported. Mentor and protégée pairs may be arranged by the organization or encouraged to form their own connections. Firms sponsor mentorship programs for the same reason they support technical coaches; these programs attract talent, help the bottom line, and demonstrate a commitment to rapid talent leadership development and retention [4].

The recommendations for an effective technical coach–protégée relationship apply equally to a mentoring relationship and include the importance of trust, communication, and clear goals. Additionally, all mentors should be attuned to signs of "internal sabotage" such as someone's backing away from challenges or self-doubt causing them to undermine their own intelligence or performance. When mentoring someone experiencing imposter syndrome, it can be helpful to provide validation, share one's own personal experiences with imposter syndrome, normalize the experience of imposter syndrome, and teach self-compassion [21].

Here, we will describe a few mentor archetypes.

New-to-Career Mentoring: The more experienced person benevolently explains the ropes to a novice. The successful mentoring of a novice can be life-changing. A sense of belonging can increase confidence. The focus of mentorship for a new hire is on their goals and development and the relationships are generally unidirectional in terms of benefit. One firm we learned about offers a rotational program to find the best mentorship pairings for a large cohort of new hires. Pioneered by a seasoned mentor, the 18-month program changes assignments every six months as a "try before you buy."

Situational Mentoring: Having the right discussion at the right time with the right person. Call it a "quick strike" approach, with primary value at opposite ends of the spectrum – either remedial or promotional. If someone committed a faux pas, or recently took on a leadership role, short-term mentoring can make a difference. An organization supporting your learning and growth shows their faith in you.

Reverse Mentoring: Somewhat the inverse of traditional mentoring; executives are mentored by less senior employees on "topics like technology, social media and current trends" [22]. Reverse mentoring benefits both individuals involved. Executives can stay aware of current trends that are important to younger employees, while the less senior participant experiences the benefit of being able to help someone else and contribute positively to the organization [22].

Group Mentoring: Groups consisting of multiple mentors and multiple mentees [23]. This increases the number of experts that each mentee has access to and increases the knowledge and skills they can acquire [22, 23]. Further, group mentorship allows experts to receive additional mentorship and guidance themselves. While the overall structure is a group format, the learning remains individual [22]. Each protégée works individually on their gaps in knowledge and A3 goals.

Peer-to-Peer Mentoring: Connections can form organically and evolve into a fluid exchange of ideas and support. Other names for these mentors are "peer thought partners" or "identity" mentors, reflecting connections with people who are in the same "box" you are in. Peer mentoring is not bounded by the confines of work. For example, peer mentoring can happen with new mom coworkers bonding and, together, figuring out how to manage their lives, in addition to bouncing technical ideas off one another.

Employee Network Groups (ENG) provide an opportunity to educate others and form a peer network. ENGs can be a safe place to compare notes, be our authentic selves, and let our hair down. Friendships and informal mentorships easily form, and psychological safety and work culture improve. As firms discover how they benefit when connections are formed, they completely support diverse groups organizing. We strongly believe everyone can find benefits from belonging to an employee network group.

Sponsors: A sponsor is someone who can help to open doors for you through introductions and opportunities. Mentors can serve as sponsors or a sponsor may be a separate relationship [4, 22]. Some of the women we spoke to commented on the role of sponsors in their lives. "What I had was sponsors," reflected Tietjen, now a professional author and speaker, on those who pushed her into speaking roles and the spotlight. Dee, a human factors engineer, used the term "advocates" to describe people who have opened doors for her. Johanna noted the necessity for mentors to advocate: "We mentor to death and do a lot of loose coaching, but we do not do the work to get them exposure, i.e., the push to get them through the door."

External Organizations: Beyond internal connections within the firm, there are also mentoring connections in external organizations, professional societies, and national and local networking groups. Brooke, a senior engineer, found all her jobs actively networking within the Society of Women Engineers. Tess, a radiation scientist, explained, "I go to national and international industry networking events." She specified that she is not looking for specific people but hands out business cards to all levels and diligently keeps in touch. The resulting knowledge exchange forms mutually beneficial relationships or networks worldwide.

E-mentoring: Remote connectivity became robust during the COVID-19 pandemic, and we have learned to talk and virtually meet with anyone from anywhere in the world. Professional society meetings, once exclusively in-person, went online during the pandemic. Social media makes it easy to reach out in a professional capacity via LinkedIn to someone else with more expertise or experience in a knowledge area [22]. Other online groups to meet other professionals include Science Career Forums and Meetups (search "meetup" followed by the field of interest).

Gender Matched versus Gender Mis-matched: Mentors play different roles in our lives depending on our career stage, goals, and identified needs [16]. We may have multiple mentors at a time who support us in different ways. However, there are times when it is most helpful to have a mentor who has shared life experiences. For women, therefore, there may be particular moments in their lives when having a female mentor is crucial [22]. For example, becoming a mom or significant caregiver can be a major career inflection point. Gayle, a scientist, recognized that now that she is a working mother, she needs a mentor who is a woman. Additionally, early-career women may have issues with self-confidence, imposter syndrome, and, in fields like engineering, working in a primarily male environment. A female mentor who has had prior similar experiences can provide wisdom and guidance in a unique way.

CONCLUSION

A good match, good listening, safety, trust, and accountability are a few of the main ingredients of supportive relationships that can facilitate sustained leadership and innovation. Through describing the importance of role models, technical coaches, and mentors, this chapter brings to the forefront examples of positive structural improvements to improve work climate, facilitate control over career, and enable satisfying work. These approaches drive inclusion, respect, and professional advancement toward gender parity, and we hope there will be widespread intentional adoption of these practices.

Present State

Parts I and II demonstrated how with few role models and with paths for advancement not clearly discernable, current work practices in engineering are failing to retain women. Further, work dissatisfaction is one of the top reasons women leave engineering. Role models, technical coaches, and mentors are valuable in supporting women in engineering through career development and creating a sense of belonging. Women have access to fewer informal mentors and technical coaches than men [4].

Promoting Change

Identify a Role Model/Be a Role Model: Look around and identify a role model, someone whom you would like to emulate who shares your values and the chosen career path you hope to pursue. Be aware, *you can be a role model* just by virtue of being a woman in STEM. You have a platform.

Find a Technical Coach and Mentor: Having technical coaches and mentors enables you to learn more effectively because you can directly share and explore knowledge gaps with an empathetic partner who cares about you and your work. Technical coaching can most easily be done through the supportive reviews of an A3 the protégée has developed for their project work.

Make It Known You Are Eager for Challenges: Continually challenge yourself to learn. Based on the discussion of gender schema in Chapters 2 and 3, we know that there is a tendency to view women as less competent than men. In line with this, we have heard reports of managers deciding not to offer women the challenging assignments and positions predicated on the belief that women are less capable and less willing to take on tougher projects or hold more authority. Make it known that you are looking forward to a challenge and learning new things.

Take Risks, Grow: Believe in yourself. Self-doubt did not get you to where you are today. Do not hesitate to go after what you want. Innovators learn through a series of failures. Several women we know have experienced big, even nearly catastrophic, failures; however, with a supportive network they were able to get right back on track. Being an engineer takes tenacity. There is no chance of triumph without attempt.

Support Other Women: This is not a zero-sum game. Johanna regularly mentors four to five people and seeks to create an easier path so they can avoid obstacles. Annette shared, "I would have liked to have more role models who were female-friendly."

Remember to Give Back: As a way of life, pass the support you received on to others. Be sure to show gratitude to those who have supported you.

A Few Words of Thanks

We would like to extend our gratitude to all the amazing women who have shared their stories with us, sometimes feeling vulnerable, being completely authentic and humble. We celebrate the journeys that we each have taken since earning our engineering degrees. Regardless of whether you remain active in engineering or have pivoted away, you have developed a rich skillset for your life's toolbox.

REFERENCES

1. M. L. Frazier, S. Fainshmidt, R. L. Klinger, A. Pezeshkan, and V. Vracheva, "Psychological Safety: A Meta-Analytic Review and Extension," *Personnel Psychology*, vol. 70, no. 1, pp. 113–165, Feb. 2017, doi: 10.1111/peps.12183.
2. T. Geraghty, "The Difference Between Trust and Psychological Safety," *Psychological Safety*, Nov. 16, 2020. https://www.psychsafety.co.uk/the-difference-between-trust-and-psychological-safety/ (accessed Dec. 28, 2021).
3. "5 Tips to Be an Effective Technical Mentor," *Ohio University Online Degree Programs*, May 10, 2021. https://onlinemasters.ohio.edu/blog/technical-mentorship (accessed Dec. 18, 2021).
4. S. Dinolfo, and J. S. Nugent, "Making Mentoring Work," Catalyst, New York, NY, 2010. Accessed: Feb. 05, 2022. [Online]. Available: https://www.catalyst.org/wp-content/uploads/2019/01/Making_Mentoring_Work.pdf.
5. Cambridge Advanced Learner's Dictionary & Thesaurus, "Role Model," *Cambridge Advanced Learner's Dictionary & Thesaurus*. Cambridge University Press, Cambridge, 2013. [Online]. Available: https://dictionary.cambridge.org/us/dictionary/english/role-model.
6. R. Gotian, "Why You Need a Role Model, Mentor, Coach and Sponsor," *Forbes*, Aug. 04, 2020. [Online]. Available: https://www.forbes.com/sites/ruthgotian/2020/08/04/why-you-need-a-role-model-mentor-coach-and-sponsor/?sh=c45fa987c489.
7. K. Cherry, "Biography of Abraham Maslow (1908–1970)," *Verywell Mind*, Mar. 16, 2020. https://www.verywellmind.com/biography-of-abraham-maslow-1908-1970-2795524 (accessed Dec. 20, 2021).
8. C. N. Winston, "An Existential-Humanistic-Positive Theory of Human Motivation," *The Humanistic Psychologist*, vol. 44, no. 2, pp. 142–163, Jun. 2016, doi: 10.1037/hum0000028.
9. T. R. Clark, *The 4 Stages of Psychological Safety: Defining the Path to Inclusion and Innovation*, First Edition. Oakland, CA: Berrett-Koehler Publishers, Inc, 2020.
10. S. Lee, S. Kwon, and J. Ahn, "The Effect of Modeling on Self-Efficacy and Flow State of Adolescent Athletes Through Role Models," *Frontiers in Psychology*, vol. 12, no. 661557, Jun. 2021, doi: 10.3389/fpsyg.2021.661557.
11. A. Bell, R. Chetty, X. Jaravel, N. Petkova, and J. Van Reenen, "Who Becomes an Inventor in America? The Importance of Exposure to Innovation," *The Quarterly Journal of Economics*, vol. 134, no. 2, pp. 647–713, May 2019, doi: 10.1093/qje/qjy028.
12. K. E. Wohlford, J. E. Lochman, and T. D. Barry, "The Relation Between Chosen Role Models and the Self-Esteem of Men and Women," *Sex Roles*, vol. 50, pp. 575–582, Apr. 2004, doi: 10.1023/B:SERS.0000023076.54504.ca.
13. B. Wilder, "When the Student is Ready, the Teacher will Appear," *Industry Week*, Oct. 31, 2013. Accessed: Dec. 23, 2021. [Online]. Available: https://www.industryweek.com/leadership/change-management/article/22009744/when-the-student-is-ready-the-teacher-will-appear.
14. J. F. Madden, "Performance-Support Bias and the Gender Pay Gap among Stockbrokers," *Gender & Society*, vol. 26, no. 3, pp. 488–518, Jun. 2012, doi: 10.1177/0891243212438546.
15. R. L. Badawy, B. A. Gazdag, J. R. Bentley, and R. L. Brouer, "Are All Impostors Created Equal? Exploring Gender Differences in the Impostor Phenomenon-Performance

Link," *Personality and Individual Differences*, vol. 131, pp. 156–163, Sep. 2018, doi: 10.1016/j.paid.2018.04.044.

16. P. McCauley Bush, *Transforming Your STEM Career Through Leadership and Innovation: Inspiration and Strategies for Women*. London: Academic Press, 2013.
17. S. Gallo, "Coaching as an Employee Benefit: It's a Win-Win," *Training Industry Magazine*, Dec. 05, 2019. Accessed: Dec. 18, 2021. [Online]. Available: https://trainingindustry.com/articles/performance-management/coaching-as-an-employee-benefit-its-a-win-win/.
18. E. Soken-Huberty, "10 Reasons Why Trust is Important," *The Important Site*. https://theimportantsite.com/10-reasons-why-trust-is-important/ (accessed Jan. 17, 2022).
19. L. Phillips-Jones, *Skills for Successful Mentoring: Competencies of Outstanding Mentors and Mentees*. Grass Valley, CA: CCC/The Mentoring Group, 2003. [Online]. Available: https://my.lerner.udel.edu/wp-content/uploads/Skills_for_Sucessful_Mentoring.pdf.
20. D. J. Dean, *Getting the Most Out of Your Mentoring Relationships: A Handbook for Women in STEM*. Dordrecht, NY: Springer New York, 2009.
21. W. B. Johnson, and D. G. Smith, "Mentoring Someone with Imposter Syndrome," *Harvard Business Review*, Feb. 22, 2019. [Online]. Available: https://hbr.org/2019/02/mentoring-someone-with-imposter-syndrome.
22. K. Zimmerman, "Modern Mentoring is the Key to Retaining Millennials," *Forbes*, Jul. 18, 2016. Accessed: Nov. 20, 2021. [Online]. Available: https://www.forbes.com/sites/kaytiezimmerman/2016/07/18/modern-mentoring-is-the-key-to-retaining-millennials/?sh=2d1fa19c5fc8.
23. B. N. Carvin, "The Hows and Whys of Group Mentoring," *Industrial Commercial Training*, Feb. 2011.

Part IV

Summary

In Part IV, we described three elements to empower control over a career.

- In Chapter 10, we identified the areas that create an unlevel playing field and introduced the concept of A3 to help level the playing field. At the center of this unlevel playing is the need to gather knowledge from peers. The second area is the relational work that women might migrate to or be encouraged to move to. This shifts them from what may be higher valued technical roles. The third area was to communicate that the primary purpose of Lean Development is to increase the learning rate of individuals and therefore grow their knowledge.
- In Chapter 11, we described the six principles of Lean Development used as the vehicle to change the rules of the game. We identified three of these principles that an individual can focus on: creating reusable knowledge through the use of A3 reports while doing the work, system thinking, and set-based design. These three elements directly attack the way the work is done and how learning occurs. These areas are focused on the mechanics of the development effort.
- Finally, in Chapter 12, Deb and Rose described professional relationships that are critical for supporting women in engineering: role models, technical

DOI: 10.1201/9781003205814-21

coaches, and mentors. Through career development and creating a sense of belonging, these relationships increase work satisfaction and retain women in engineering.

The methods of Lean Development coupled with the professional relationships described in Chapter 12 are the first part of our change to the design of the engineering system. As we move into Part V, we will now shift our focus to enabling the creation of satisfying work.

Part V

Strategy 2

Enabling Leaders to Lead – Creating Satisfying Work

13 Lead the Change to the Way the Work Is Done

Leadership theory abounds in many areas but, for our problem, let's consider two elements: recognizing the need for change and establishing the strategic insight into what to change and how to do it. In Part IV, we identified how to implement the shift needed for someone to control their career, given a basic level of management support. As we move into Part V, we now describe the second half of our strategic direction: organizational change that builds a culture of learning and leadership consistent with Lean Development. The ultimate goal is to create satisfying work. We are using Lean Development as a foundation for that change with the knowledge that individuals benefit from this effort, both personally and in relation to their careers. A business also benefits because of increased efficiency, improved profit and revenue, more innovation, and higher retention of female engineers.

THE LEADER'S ROLE AS A DESIGNER

In Peter Senge's book *The Fifth Discipline* (1990), he writes in the book's section titled "Leader as Designer" [1, p. 341] how he has asked many group managers to envision their organization as an ocean liner and, to consider, as "the leader" of that ship what their job would be [1]. The most frequent answer he received was that managers picture themselves as "the captain." Others might say "the navigator." However, the role that is generally neglected is that of the "designer of the ship" [1, p. 341], the role that defines how the ship can operate. Senge goes on to give the example that even if the ship's captain calls out for a starboard turn, if the ship's designer did not enable that capability in the function of the ship, it cannot do a starboard turn.

Just like an ocean liner, the leader as a designer is critical to reaching our fundamental objective of designing the engineering system to enable women's success and ultimately the organization's success.

We begin our work in Part V drawing upon the six principles of Lean Development that we described in Chapter 11. In Chapter 11, we suggested that these principles, beginning with reusable knowledge, allow us to change the way the work is done through the work of a System Designer and Set-Based Design. In addition, we suggested that the first three principles contribute to building the foundational processes of an organization while the last three are the principles that deliver value to the business.

DOI: 10.1201/9781003205814-23

WORKING AS SYSTEM DESIGNER AND MOVING THE ORGANIZATION TO A3s

From our seafaring story in Chapter 6 and then the work in Chapters 10 and 11, describing the A3 Problem-Solving Process methods, we have established that our navigational clock is an A3. The A3 report can facilitate greater success for an engineer, just like an accurate, reliable, clock can be in the hands of a ship's navigator.

A3s empower managers and others to coach individuals. By creating A3s, this is what we are working toward.

DOING WORK BY A3

A new-to-career engineer is showing her work to her peers through an A3 report that she has started. Using the sheet, she is working to clearly communicate what she is trying to accomplish, how she is doing it, what she has learned, and knowledge gaps. Through individual or small group reviews of her sheet, she is learning how to communicate its purpose and intent. Those who are helping with her learning and decision-making see what she has put on paper and, where the content is not clear, have a firsthand opportunity to help her find that clarity. They help her find areas of focus that have been missed in her work. Any implicit bias in those reviews is mitigated because they see what she has written.

Each review of her sheet is a learning opportunity for her and a teaching opportunity for the people on the other side of the desk. During those reviews, she is not only moving her technical or business knowledge forward, but she is also seeing how someone can use the Socratic method of asking thought-provoking questions to create new ideas and teach someone how to think. As she shares her sheet with a broad set of people who have a diverse set of knowledge, experiences, and perspectives, she naturally builds and broadens her breadth of knowledge. In addition, she is becoming a system designer, developing system-level thinking skills in the process. She is learning by doing and becoming an expert in this area of work. She is learning how to take control of her work and create satisfying work. Through this effort, she learns how to learn and builds her own confidence.

The sheet eventually becomes a holistic view of the work she has done. When an individual in another group needs to understand her work and the resulting solution, they send her an e-mail to get a copy of her A3. The engineer replies with a copy and points them to the location where it is stored on the department's shared drive.

Looking back at the graph we created in Chapter 10, Knowledge Acquired versus Time, she has moved herself to the 45-degree line (Exercise 10.1).

So with that being our ultimate goal, let's look at where we are now by playing out the same activity above through a typical organizational approach. Without being told of another method, the individuals involved will continue to do the work the same way that the organization has relied on for years.

DOING WORK BY SLIDEWARE

The new-to-career engineer is putting together a slide deck to be used as part of a presentation of her design work to her team and others involved in the project. She reviews her "slide deck" with those who want to help her. She is working to gain their insights, perspectives, and knowledge; however, some of those reviews are more focused on how to best sequence the slides to "bring the audience along" so as not to overwhelm them. There may be discussions about whether she should identify issues at the beginning of the slide deck, so there is clarity about the challenges, or near the end after she has communicated the situation. She may be told that some managers want it one way while others want it the opposite way. However, she is left with no good choice in this case considering both types of managers are in the same review audience.

The discussions of her slides will include the basic presentation challenge of finding a good balance of what is shown on the screen versus what is verbalized. So, she decides to work toward the basic approach that a good presentation has 50% of the meeting content provided from the slide deck while the remaining 50% is verbalized. With these individual reviews, it is unlikely that the discussion is framed from a perspective of teaching her how to think or write concisely with clarity. After all, it is just a slide deck for a presentation that, at times, feels like just one more check box in the process of the development work. In addition, these reviews start with the goal of approving or supporting the work she is responsible for to ensure that the ultimate decision-maker supports her recommendations and the work she has done.

In the broad team review of her slides, she may see a variety of interest levels in the content, as demonstrated by the number of people engaged in her presentation versus doing their own tasks on their PCs. In addition, given that she is new to her career, she may find it challenging to verbally communicate nuances of the problem or solution, particularly if having to speak over opinions from those in the room. Meanwhile, this team review will generally fall on the side of presenting the content rather than gathering knowledge from others. Her three recommendations on Slide 20 are accepted. However, the slide lacks technical depth because that was presented ten slides earlier. Given that she put effort into the presentation and worked with a number of people in its development, she more than likely did a good job and would be told that at the end of the review. Sadly though, whatever does make its way into the slide deck will most likely end up as work that is e-mailed to the team and archived somewhere with the minimal expectation that others use it. Meanwhile, by design, the slide deck is only 50% of the information, the other 50% just walked out the door at the end of the meeting.

As we look at the use of A3s as a tool to drive change, we can begin with the perspective that their use and the use of a slide deck are both physical tools that can be measured against each other. We can then compare how each one is used in the course of the problem-solving process and, fundamentally, the work without evaluating how discussions happened, who might have talked over who, or whether the conversations had bias throughout the discussion. We can look at the content on each of them and ask the questions: What amount of technical or business content does each contain? How do they actually move the development work forward? But just as importantly, as seen in our example above, if a slide deck contains only 50% of the information, bias can clearly drive the other 50%.

Using the methods of Lean Development, the leader has the opportunity to teach her organization and the people within it how to learn. The leader has the chance to increase the rate of learning for the individuals and ultimately the organization. Most importantly, a leader can level the playing field for women and empower individuals to maximize their contributions. A3s are a tool to do this.

However, just having engineers and managers create A3s as part of doing the work is not sufficient. The true benefit comes from adding structure into the work, which requires sharing the A3s with others. This brings us to the last piece of our ocean travel analogy from Chapter 6 and the use of a time-ball to ensure that a ship's navigator clock had the correct time.

We can imagine that when the ship is moored in port, the ship's Captain might expect that the ship's navigator is setting his clock from the port's time-ball and recognizing how much deviation is occurring. Once at sea, the navigator's understanding of the accuracy of his clock will establish how well he can navigate. This same effect occurs when the engineer regularly shows her A3s to others. Regular reviews build knowledge, capability, and confidence, as well as improve accuracy. Meanwhile, the leader sets the tone for this method of work. This in turn creates a learning culture, which effectively moves people and the organization to the right on the x-axis of Figure 9.2.

A3 REVIEWS ARE OUR TIME-BALL

Going back to our history lesson on finding longitude (from Chapter 6), the regular review of an A3 by managers and others is our time-ball. These regular reviews, based on a schedule to meet the needs of the author and the ultimate project objectives, establish the cadence of learning. We are leveraging the second principle of Lean Development, Cadence, Pull, and Flow, to drive the rate of learning as we implement the third principle of Visual Management.

IMPLEMENTATION OF LEAN DEVELOPMENT AT THE DEPARTMENT LEVEL

We approach change management with the following basic elements:

- It is always easier to change the process than it is to change behaviors, knowing that as the process is changed the desired behaviors will follow – when those desired behaviors are clearly communicated.
- Change requires both top-down and bottom-up efforts. The top-down change is leading by example and the bottom-up change is showing the management team what works and what doesn't work. When the engineering level utilizes A3s as part of doing their work, they are demonstrating bottom-up leadership.
- We will focus first on four areas: the use of A3s, establishing a knowledge depository, the technical depth of managers, and developing systems thinking through system design.

1) The Use of A3s

The first area of focus, and the key to the entire change in process, is to wean both the management team and engineers from the use of slide deck presentations and replace them with A3s. The primary objective is to improve the problem-solving process while creating reusable knowledge. A departmental focus would be the creation of an A3 that captures the firm's product portfolio* to clarify business goals/objectives and individual product objectives. This sheet can establish initial customer needs and manage product investments. Creating an A3 Product Portfolio leads by example. The department manager needs to ensure that this work, done at her level, is actually helping the engineers do their work better and faster. What is done at the department level must be directly transferable to the project and engineer level.

Based on almost two decades of their use outside of Toyota, we know that the use of A3s can be successful. However, there are more than a few case studies in which an organization goes down the A3 path and ends up with many A3s and little business benefit. So, this area needs to be deliberately planned and managed well. If it looks like their use is not working, it is an opportunity to evaluate what is going wrong. This is no different from any other development activity.

* Product portfolio integrates all the products or services that a firm would offer to customers.

2) Creating a Knowledge Repository

The second area of focus is to identify a simple way to capture the knowledge created via A3 reports, support reusable knowledge, and make it visible to all individuals. Each program should have a place to keep knowledge, and each department should have some form of a central knowledge repository. The key is doing something that is simple and easy. If someone is looking for a specific A3, they should be pointed to the location. Mixing A3s with past documents that have been created over the years may create challenges; sometimes it is easiest and best to start with a clean slate. In addition, deciding to implement a new file management system can create a level of complication and defocus that may not be helpful. The key is to find a simple workable solution to keep the specific files of A3s and their supporting documents in a location where people can be directed to.

3) Technical Depth of Managers

The third area of focus is the technical depth of managers. Our causal diagram, Figure 6.2, showed how technical capability feeds into personal growth and business contribution. This is true for both engineers and managers. Many firms have implemented a dual-career path that includes a technical leader path and a management path. This has been beneficial in building and retaining the technical depth for individuals who don't want to pursue the management path; however, in some cases at the cost of removing the need or expectation for the managers to maintain or build the depth of their technical skills.* The technical career path still needs to be maintained and Lean Development will accelerate that growth; however, increasing the management team's technical depth is critical to fully utilizing Lean Development. Building or maintaining the technical capabilities of managers is easily done through the A3 process. As managers are reviewing A3s of their direct reports, they see the technical work being done or to be done and have the opportunity to ask thoughtful questions about the work. During those reviews, they can focus on the areas that have technical challenges, thus building their own knowledge. They are given a firsthand opportunity to learn while at the same time coaching.

The act of managers reviewing and contributing to the A3s created by engineers and other managers, although beneficial to all, is most beneficial to the woman manager as it helps to level the playing field. This is shown in the causal diagram (Figure 6.2) in which reusable knowledge drives four aspects of the work and environment:

- Increasing Confidence, in the long-term, can ultimately decrease Bias (see Figure 2.1)
- Improving the abilities of women Role Models
- Increasing one's Control over Learning
- Building Technical depth which leads to Satisfying Work

* Camille Fournier has a great book titled *The Managers Path* (O'Reilly® 2017) that stresses the importance of managers staying technically relevant, starting on page 153 [2].

Coaching in Set-Based Design (which includes the use of trade-off curves), and encouraging the use of it, should be a component of increasing the technical depth of managers. The manager plays a key role in helping to establish the development activities, prioritizing resources and the decision-making processes. Therefore, the manager is a key leader in helping to move the organization away from Build–Test–Fix and toward Set-Based Design. This work can identify the knowledge gaps that need to be addressed. So, rather than a manager just looking to an engineer for a recommendation between two designs, the manager is now in a position to talk with the engineer about the technical trade-offs between the two designs using the trade-off curves.

4) Development of Systems Thinking through the System Designer

As we covered in Chapter 11, the role of a system designer is a critical principle of Lean Development. As we begin to move the organization toward Lean Development, we want to be overt about developing systems' thinking skills. We want to help people become system designers from the department level all the way to the engineering level. As we describe this process, we are using the term system designer in the general sense. The individual naturally looks at the entire system they are responsible for as a normal part of their work. When doing their work, they are able to integrate across multiple objectives. In contrast, a system designer can map out the flow of the development work from the very beginning of the work to the end. In addition to executing their work from a system perspective, they can coach others in the area of systems thinking. This is done by coupling the efforts of creating system designers with the use of the A3 process. When the system is put down on paper, it can be seen as a whole.

These four areas of focus are all core to Lean Development and form the basis for the first efforts to change the system of engineering. They are grounded in changing the way the work is done and are all easily measured. They are also directly linked to improved organizational performance based on the research and case studies in Lean Development. This top-down work using A3s can happen in a matter of weeks, not months. Each of these areas of focus is then executed at the Project Team and Engineering levels to enable bottom-up change.

IMPLEMENTATION AT THE PROJECT LEVEL

Within the top-down implementation of the A3 process, there are three levels of work, as shown in Figure 13.1.

1) The department's Portfolio A3 flows down to the project teams.
2) From Portfolio A3, an individual project team manager leads and owns the development of the Project Level A3, which pulls the goals, objectives, budget, customer need, and so on, from Portfolio A3 to Project A3. The creation of Project A3 may bring up questions about the Product Portfolio, which are then brought up in a discussion about Portfolio A3 and may

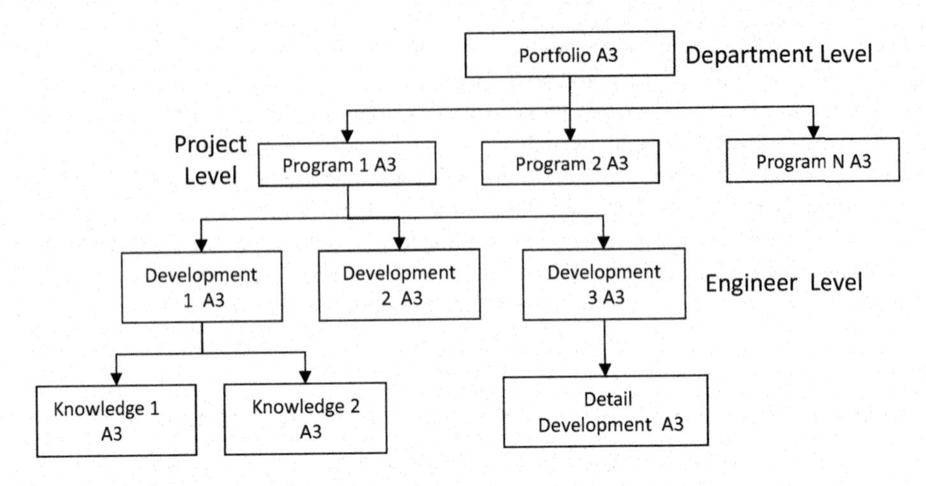

FIGURE 13.1 The cascading structure of A3s.

lead to revisions. As a result, there is an iterative loop that builds a common understanding among all those responsible for the project-level work. During this phase, the project lead works with her design team to establish the development strategies, design the flow of work, and identify the knowledge gaps.

3) The project team lead then works with individual engineers, at all levels of experience, to help them create A3s for the development work they are responsible for. These sheets draw from the Project Level A3 and will be the primary tool that an engineer uses to map out their work, gather input from peers and more senior engineers, and receive coaching from the project lead and others.

In general, the complexity of an A3 is driven by the complexity or amount of work that is required to meet the goals identified on the A3. As a starting point for the Project Team Lead, it may be helpful to just assume that every engineer owns the creation of at least one A3.

IMPLEMENTATION AT THE ENGINEER LEVEL

With the support of the project team lead, the engineers start their work, drawing from their specific Program A3. They work through the creation of Development A3, which outlines the objectives (left side of the page) and the specific development work on the remainder of the sheet. Based on that Development sheet, the work may require specific Knowledge A3s focused on the creation of detailed technical knowledge. The work on either or both of these sheets would include the required knowledge to be obtained with a focus on Build–Measure–Learn and, specifically, Test-Before-Design. This is where our use of Set-Based Design shown in Figure 11.1 drives the order of the specific learning. The engineer reviews these sheets with project leads and others through a review and learning process to progress the work forward. At the beginning of this chapter, we would expect this to happen in the story, "Doing Work by A3." Just like with Project A3 and Portfolio A3, the review of Knowledge A3s and Development A3s may result in changes to Project A3 and subsequent modifications to Portfolio A3. This iterative approach ensures that all members of the team are on the same page and that there is open communication. It also reduces the potential for errors or mistakes, facilitates creative thinking, and generates new ideas.

The key point in this A3 structure is that the sheets, although just a tool, help guide the development work and create a learning process within the control of the engineer.

WORK–LIFE BALANCE: WORKING PART TIME AND GOING HOME ON TIME

With our implementation of Lean Development and the use of A3s beginning at the department level and then moving down to project management and then to the engineering level, we can now take advantage of that effort and apply it to improving the work climate, specifically work–life balance.

Working Part Time

As we saw noted in Chapter 5, one study found that 27% of women with engineering degrees left engineering and the majority of those left due to their inability to find part-time work [3]. Based on this statistic, designing the engineering system to accommodate part-time work could have a significant impact on retaining women engineers. In addition, drawing upon our use of medicine as a case study, we also know that women physicians have the ability to work part time, with one study finding that 22.6% of a cohort of early career women physicians were working part time [4]. As we look at the design of our engineering process, we want to understand how to create this level of flexibility within an engineer's career.

Looking at the work of an individual, we can break it down into three very basic activities:

1. Establishing the goals or objectives of the work
2. Establishing and aligning the work with other activities and the work of other individuals
3. Conducting the work to completion while meeting the objectives

Within each of these steps, a myriad of activities, skills, and knowledge are required in order for them to be successfully completed. While we could certainly break down these steps to a more detailed level, at this general level, the engineering process can be managed in a repeatable and predictable manner when the system of engineering allows for it.

Therefore, we can use two steps to allow the process to accommodate part-time work:

1. The leadership team sets the expectation and tone that a portion of the engineering work can accommodate part-time work without compromising an individual's career.*
2. The methods of Lean Development are used to create opportunities for part-time work.

So how might this look?

It begins with the management team clarifying the organization's work and workflow process. This can be done during the process of creating Program A3s as shown in Figure 13.1. While making the A3s, managers can either establish areas of work that can accommodate part-time work or design the process of work in order to accommodate part-time work. Basically, they can establish what in the development work can be done on a schedule that is not 40 hours a week. Then, through an engineer's use of Development A3s and Knowledge A3s (Figure 13.1), the engineer manages their own approach in going through the three steps listed above. Using these A3s, a manager can facilitate any unique work needed during the off-time of an individual, not from the perspective of helping out but as part of the defined development process. In other words, the overall development process is designed to meet the cadence of a five-day, 40-hour work week but accommodates pieces of the work functioning at less than that, for example, three full days a week or five half days a week.

In some cases, creating opportunities for part-time work may need to be accomplished in the form of job sharing. Not necessarily with two individuals both working part time and sharing one full-time job, but through a part-time individual and a full-time individual sharing specific development work efficiently and effectively. The key element in the use of this approach is that the A3s created to do the work enable individuals to more effectively work together. In addition, this may be a method to allow managers to work part time as well.

Going Home on Time

Aside from facilitating changes to the design of the engineering system to create part-time work opportunities, we know that the implementation of Lean Development enables someone to do the work better and faster. The work is faster because the learning is faster. For the person that feels that working part time is their only option to reduce the impact of long work days, we can focus on a very simple objective of "going home on time." This effort is intended to address the 40-hour work week that regularly turns into 50 hours, with little control over those additional hours.

* The reality of the situation in which someone works half-time for two years is that they will see half of the opportunity for growth and learning. However, it should not change the trajectory of the desired career path, just delay it.

Meeting this simple objective may meet the needs of many people looking for work–life balance, which should include everyone, but in this case, we are focusing on this effort in order to increase the retention of women in engineering.

For a more detailed approach to the objective of "going home on time" through eliminating waste (non-value-add work), see Appendix A and specifically Figure A.1. The general approach begins with understanding that there are three types of work: Value Add, Non-Value Add, and Non-Value Add but Necessary. The next step is to take a deliberate and personal step and categorize the different components of work being done into one of the three buckets. Then, apply the methods and tools of Lean Development to overtly eliminate Non-Value Add work.

Within a progressive firm, this need or desire to enable part-time work or work–life balance is not new. However, what might be new is the idea that the design of the engineering system using the principles of Lean Development can make it possible. Importantly, it needs to be implemented in a way that prevents someone moving to part-time work from taking a step backward in their career. In other words, they are not forced to move to another organization that may not be their desired choice in order to work part time. Or, worse yet, have it appear that their commitment to their career is somehow lessened because of working part time.

MANAGING THE WORK

The four primary A3s being used, Portfolio, Program, Development, and Knowledge, permit a design area to manage the work. The use of Development and Knowledge A3s by engineers will start to create a culture in which technical discussions are centered on what is written down and can be debated in technical discussions. This will reduce the level of opinion-based design reviews that may occur using an administrative slide deck. As each engineer creates her own Development and Knowledge A3s, a department is able to build a knowledge base that is available to all within it. Each individual engineer has the opportunity to learn from others through the A3s created. Suppose an engineer is interested in moving into a new technical area. In that case, they have the opportunity to develop an initial understanding of that work by reviewing the available knowledge, or A3s, created by others. Because each of these sheets has been created to be "A" level work (just like in school), the sheets can basically create a college-level textbook.

Let's look at a hypothetical example demonstrating how the work is done and its effects.

PART I. THE WORK ASSIGNMENT, FROM FIGURE 2.1 – GENDER CONFIDENCE AND BIAS

A manager has two relatively new-to-career engineers, one man and one woman – both new to his group. On paper and based on their experiences and what the manager knows about them, they are identical in skills and capabilities. The manager has two assignments: one difficult and one less difficult. In talking with the male engineer, the engineer expresses confidence in his ability to take on the difficult assignment; however, when talking with the woman engineer, she is less confident. (We know that this is a reasonable situation based on our understanding of the gender confidence gap.)

Either because of the manager's bias or an inherent desire to make sure nothing gets in the way of getting the job done and the success of the engineers, the manager elects to give the more difficult task to the male engineer. After the manager gives each of them their work assignments, the manager asks the engineers to go off and do some initial work on their assignments and come back in a few days to talk about what they have learned about it and how they will approach the work.

As we look at this example through the view of our Implicit Bias – Confidence Model (Figure 2.1), an individual's lack of confidence can impact the manager's choice of work assignment which can have external and internal effects. On an external level, it results in a less difficult work assignment for the woman. On an internal level, if the manager says, "I can appreciate your reservation about taking on the more difficult assignment, so maybe it is best if we have you work on the less difficult one," the woman engineer's confidence may be further reduced as her beliefs about her low abilities feel confirmed. In our story, it is also possible that because of his or her own biases, the manager may have consciously or unconsciously described the two projects in a way that steered the female engineer toward the less difficult assignment and the male engineer toward the more challenging assignment. The downstream effect of this is that the male engineer's confidence may further increase through this process while the female engineer's confidence decreases. Consequently, the gender confidence gap widens further.

Additionally, the manager's choice of assignments has a much broader impact than just what two engineers might think of themselves or the experiences they may gain. In fact, given that they are both early-to-career engineers, the actual work they are doing may be of a low level of importance to the overall group's objectives and challenges. However, the most concerning issue is that their work assignments are visible to the entire group. This simple work assignment choice may perpetuate the belief that men work on the more difficult work while women pick up the rest. The choice in assignments, which may be a simple choice for the manager to make, reinforces underlying biases and demonstrates to the other women within the group the type of work they should expect to do. Even though it was never set up as a competition, this situation created a competition between the two engineers and the woman lost because nothing in the design of the engineering system prevented it from occurring. Given the systemic nature of our gender schema, a systems-based approach is required to address the problem.

In the second part of our story, we are going to compare the methods and approaches that each of the engineers chose to use to complete the initial work. However, we need to start by explicitly describing the ways that bias may have occurred in that first interaction. As we discussed in Chapter 2, it is reasonable that the woman might identify different types of bias directed at her, all possibly unknown to the manager. However, because the female engineer understands that these have occurred, she elects to approach the work differently than what might be expected. In other words, she understands Figure 2.1 and is going to address "how the work is done" in order to counteract the bias that might have existed. Because the organization is using Lean Development methods to create a system of engineering that delivers better results, her focus on approaching the work differently is what would be expected in the situation – although, at this point in the process of implementing Lean Development, the manager may not have a similar view.

PART II. THE ENGINEER'S INITIAL WORK

A typical approach to the initial work from Part I of our story would be for the engineer to investigate the work, talk with peers, identify a plan, and then meet with the manager. In that discussion, the engineer would talk about the work from "a notebook," which would naturally only have a portion of what someone knows because the rest of the knowledge is "in the engineers head." Because the engineer prepared for the meeting, the engineer would do a good job of articulating the situation. However, it is still just a verbal discussion and the nuances that exist may not be fully internalized by either the engineer or the manager. What is best shown in a diagram is instead verbally described. Meanwhile, the manager is taking notes in his notebook. At the end of the meeting, the manager has a general understanding of the work to be done and feels good about the plan the engineer developed. The engineer followed the normal approach of doing the work and did a good job. In our example, this is the path the male engineer chose.

However, the female engineer decided to do the work differently and used an A3. After getting the assignment, she spent time creating her A3. She starts some general discussions with her peers about her sheet. She works to put almost everything she has learned about the work and what she will do on the sheet. Her sheet is a structured and clean approach using the A3 process. When she arrives for the meeting with her manager, she has two sheets, one for her and one for her manager to review. At this point, the sheet might be 25% full. Her sheet focuses on clarifying the problem, the goal, the situation, and her plan. As the manager reviews her sheet, needing very little verbal explanation, he develops a clear understanding of her work and, most importantly, the work she can do. Whatever bias that might have existed is virtually eliminated because he is seeing what she has written in a technical and clear manner. In addition, because the engineer is working to get an "A" on the final sheet, she will probably hear the manager say, "this is a really good start, send me an updated version and let me know when you want me to look at it next." To be clear, this method is not extra work, it is the work that delivers better and faster results. In fact, we can assume that the conversation was completed in about half the time as the verbal discussion we described above.

In our example, the female engineer elected to do the work differently. She took responsibility for writing down how she was going to do her work, not as a summary or slide deck presentation but in a form that helped her do the work better and faster. In addition, based on the discussion in the review meeting with her manager, we would expect her confidence to increase. The increased confidence, which may have reduced the bias, maybe in small steps, but things build on each other when they become the norm for doing the work.

Finally, we might expect that the next time there is a challenging opportunity for a female engineer, who still may lack full confidence, the manager would ask her to start an A3. The two of them can then see how things look during that first review of her sheet. In addition, the manager now has the use of the A3 as a coaching tool. Although the task may be challenging, the engineer has the tool to do the work.

USING AN A3 REPORT IS LIKE WEARING A SOFTBALL GLOVE

Our story above is a lot like watching a softball team play the game without gloves. However, once everyone on the team has a glove, the entire game changes. As a player, when a line drive is headed directly at you or you are running toward that high fly ball, things look a lot different when you have a glove on your hand than when you don't. A glove lets you feel completely confident in yelling "I got it."

When a manager expects the work to be done this way, they are changing the culture of the organization. They are moving the organization from one in which dialogue and your ability to verbally justify your opinions drive behavior to one written on paper for people to test the technical or business decisions.

Individuals and teams maximize their contributions to the business by having a clear understanding of the customer and objectives, by having the work integrated across a broad spectrum of views, by having a clear plan of how the work will be done, and by having a clear understanding of what we need to learn in order to achieve the objective.

We won't play out the example above using other possible solutions, like the woman elected to act more confidently and was then negatively viewed as "pushy" or choosing other methods of preparing for the meeting, because we know how that ends up. Specifically, no positive change for her and, unfortunately, possibly reinforcement of any underlying bias that the manager may have.

REPLAY THE DESIGN REVIEW VIA LEAN DEVELOPMENT

At the beginning of Chapter 9, we worked through how a woman engineer would approach her design work in a typical development process leading up to a design review. We also demonstrated how this example would not likely result in the woman engineer feeling empowered with the work she had done.

With our A3 process now in place, let's replay that story from Chapter 9.

A REPLAY OF "PRODUCT DEVELOPMENT IN THE TRADITIONAL WAY," PRODUCT DEVELOPMENT USING LEAN DEVELOPMENT

With the implementation of the Lean Development methods described above, the new engineer is now using the creation of her Development A3 and her Knowledge A3s to capture knowledge and establish the needed work through her interactions with other engineers and her manager. This work is directly driven by the Program A3 that her manager owns. She and her manager worked together to structure her sheets in support of the program objectives. She is capturing her Build–Measure–Learn plans and the knowledge she is gaining on her sheets. Her A3s provide a clear and concise view of the design choices that she has identified. Through the use of Set-Based Design, she has worked toward finding the strongest solution and captured the performance curves of those options on her A3. Through this work, she is becoming the expert on this particular aspect of work being done. The sheet is kept up to date, shows the state of her work, and can be shared with others at any time.

When she works with others, the sheets are at the center of the discussions. With the sheet capturing her learnings, everyone she is working with can see the same information. If there are conflicting opinions among more senior-level engineers, the sheet can be used to find and address the conflict. However, most importantly, she can feel confident in her work because her feedback from others is based on the content, work, and thought she has put into the knowledge creation and design considerations. At some point, the work will progress to the point where she is in need of a formal "knowledge review" or a major decision point – replacing the typical design review that we saw in Chapter 9.

Now let's move to the knowledge review.

A REPLAY OF "THE DESIGN REVIEW," THE KNOWLEDGE REVIEW

The "knowledge" review or decision point includes the individuals she has been working with throughout the learning process and individuals who she has not directly worked with in this effort. This review would start with a quick review of Project A3 so the broader project objectives and strategies are understood and she can show how her work directly links to the project work. She would review her Development Plan A3 so the group sees how the overall work is being approached and where knowledge gaps existed. Her Knowledge A3 is at the center of the review with a printed sheet for everyone – there is no slide deck presentation. However, she may also be showing her A3 on the projector screen.

If there are missing elements or considerations, or misevaluated risks on her sheet, they are talked through based on the information and data presented on the sheet. The conflicting views of more senior engineers have already been addressed and they would have already come to a common understanding. If individuals question the knowledge results based on their own experience, those can be identified and addressed. The most important piece here is that the meeting is focused on the last 20% of the work, the most difficult, because the first 80% of the work was already done through the process of creating the A3s. Because of this, the meeting is focused and efficient. In the end, there may still be a few things needing to be finalized, but we would expect that the work continues to the next step, design work using the strongest design option, which has been fully characterized through the trade-off curves.

After the knowledge review, she updates her A3s based on the input. When her knowledge sheet is fully complete and worthy of an "A" letter grade, it is put in the department's knowledge repository for others to use in the future. Depending on the work to be done next, she may need to start an additional A3 to manage that work. Meanwhile, she is using her Development A3 all the way through the development and implementation of her design

CONCLUSION

When leadership focuses on the four areas of shifting the organization to the use of A3s, creating a knowledge repository, increasing the technical depth of managers, and developing system thinking, they are creating structural as well as social change within the development process. A3s are at the center of this change effort because developing reusable knowledge is core to Lean Development and is a root cause of the low retention of women in engineering. Using A3s in combination with regular reviews of the A3 sheets by managers accelerates the rate of learning. The A3 process is a visible measurement of whether the organization is moving forward in the change process and the sheets need to be at the center of the discussion or, more physically, in the center of the table. Ultimately, the A3 process in conjunction with developing system designers, managers becoming technically focused, and the use of Set-Based Design, changes the way the work is done.

REFERENCES

1. P. M. Senge, *The Fifth Discipline: The Art and Practice of the Learning Grganization*, First Edition. New York: Doubleday/Currency, 1990.
2. C. Fournier, *The Manager's Path: A Guide for Tech Leaders Navigating Growth and Change*, First Edition. Sebastopol, CA: O'Reilly Media, Inc, 2017.
3. N. A. Fouad, W.-H. Chang, M. Wan, and R. Singh, "Women's Reasons for Leaving the Engineering Field," *Frontiers in Psychology*, vol. 8, pp. 875, Jun. 2017, doi: 10.3389/fpsyg.2017.00875.
4. E. Frank, Z. Zhao, S. Sen, and C. Guille, "Gender Disparities in Work and Parental Status Among Early Career Physicians," *JAMA Network Open*, vol. 2, no. 8, pp. e198340, Aug. 2019, doi: 10.1001/jamanetworkopen.2019.8340.

14 Leading the Development of Engineers and Managers

In Chapter 13, we focused on changing the way the work is done by creating reusable knowledge that builds available knowledge and confidence using a top-down and bottom-up approach. This knowledge available to all allows for the implementation of Lean Development and targets a root cause of the retention of women engineers by leveling the playing field. In this chapter, we will expand on this organization-level effort by focusing on the personal development of managers and engineers. While ultimately this looks at the personal development of individuals, we are fundamentally building a culture and a work environment that is more desirable for women and will therefore improve the retention of women in engineering. This has the overall benefit of improving the work environment for everyone. As we look at this area of leadership, we need to keep in mind that individuals have the responsibility for creating a personal development plan that helps them to achieve their professional goals. However, the management team needs to take responsibility for putting in place a structure to enable this part of an individual's effort. Managers need to understand what part of the organizational climate or work methods needs improvement or is designed to enable the success of an individual.

DRIVE LEARNING ON A DAILY BASIS

In the previous chapter, we began by describing the critical role of the designer in our needed change to the engineering system. In this chapter, we begin with the need for daily learning. We use daily, because anything else seems to disable or, at minimum, set the bar pretty low. Imagine if focused learning occurred on a weekly basis; as an engineer, who can imagine learning only once a week. To facilitate this learning, we want to go back to the foundational elements of our change, the six principles of Lean Development, and specifically the area of coaching, which facilitates learning.

Although we have focused primarily on the three principles of reusable knowledge, system designer, and set-based design, and it is very easy to see how the review of A3s is a coaching opportunity, each of the six principles provides an opportunity for coaching. Similarly, each interaction with an individual can be viewed as a coaching opportunity. For the successful implementation of Lean Development throughout the organization, it is important that the coaching provided is overtly placed within the context of the six principles of Lean Development. In other words, each manager or coach should have a mental checklist of the six principles and, each

DOI: 10.1201/9781003205814-24

time they are coaching an individual, should draw upon the principle that might be most beneficial to that particular person at that particular time in that particular situation. The six principles provide a playbook for leaders and allow managers and coaches to take a deliberate approach to the professional development of the individuals they are coaching.

A second piece of the coaching opportunity is linked to the sixth principle, specifically the function of a team. This begins with the leader looking at what elements of Lean Development should be implemented for the teams' overall benefit. For example, an engineer is asked to develop a technical design method that is a common need for a number of engineers on the team – this would of course be outlined on an A3 and readily available to the team. Another example would be implementing structured team reviews of A3s where the team that is working toward a common integration event reviews all of the A3s created as part of the development work together. In Appendix A, the "Principle 4: Entrepreneur System Designer" section, we cover this example in detail (see Figure A.4).

To understand how the implementation of Lean Development enables coaching, let's first look at a hypothetical example of someone trying to control their career in a typical organization in which roughly 20% of the learning comes from reusable knowledge and the remaining 80% of the learning comes from others or trial and error.

LEARN BY INTERVIEWING

During a one-on-one meeting, an engineer tells her manager that she would like to learn about an area of development work done by another group. She adds that she feels this knowledge would broaden her technical skills for her current work as well as prepare her for future jobs. The manager replies, "That is great! We could use more people with that knowledge."

The manager then tells the engineer that the best way to learn about that development area is to set up interviews with a few key people on that team who have spent years doing the work. This will allow her to develop a good understanding of the technical work. The manager continues to describe how she should keep in mind that the team is really busy so she will need to be aware of how much time she might be taking away from their work, or, for that matter, from being able to go home on time.

What the manager is not saying is that everything the business needs to run that area of development is in the brains of the team members. They have a lot of requests for this information and some may or may not want to share what they know with others. However, the only way for someone to learn it is to find team members that can make the time to share it.

Now, given this situation and being a female engineer, the engineer must also deal with all the challenges of how those interviews might go and what level of support she might get. She may see the influence of bias and her own confidence front and center. In addition, she may end up choosing engineers to speak to who she feels will have the most respect for her questions rather than those who might have the best understanding of the work.

Now, let's work through that same example in an organization that has implemented the principles of Lean Development – specifically an organization that uses A3 reports, recognizes its role in helping individuals create reusable knowledge, and has a knowledge repository for each development group.

LEARN BY READING

This time during the one-on-one meeting, the engineer expressed the same desire to learn about a new area of development work, and the manager's reply was still positive. However, the manager continues very differently because of the new work environment and says,

> To develop an understanding of that work, you need to start by going to the team's knowledge repository, which is set up similar to ours. In the repository, you can find all of their A3 reports. These sheets are used to map out their development work. You will want to look at the sheets they developed that were used to integrate with our team's work for the recent program you were on. Look for where your responsibilities overlapped with their work. Their trade-off curve sheets should give you an understanding of the design trade-offs they are typically working with.

The manager continues to describe to the engineer how she should identify a few sheets that seem the most relevant to her. Then, she should print them out and set up a half-hour with the team's manager to have him explain the nuances of the sheets. During that meeting, in addition to reviewing the sheets she has brought, he will most likely point her to other sheets that might be helpful. In an hour or so, the engineer should understand how the team approaches the work and what might be important for her to learn more about. The manager closes the meeting with a request for the engineer to meet with her, after meeting with the team's manager, to share what she has learned so they can talk about the next steps.

In addition, we can't help but expect that the engineer will ask a thoughtful question during the meeting with the team's manager, as they are reviewing A3s, which might cause him to think about an area of his team's design responsibility in a new way.

In the first example, the manager had little opportunity to coach other than how to approach the individual meetings. However, in the second example, learning by reading, the manager is now in a position to offer a much higher level of coaching.

Specifically coaching builds off the methods of Lean Development. The manager was able to direct the engineer to look at the A3s in very specific ways. With respect to the personal effort of the engineer to control her career, in the first situation of our hypothetical example, "Learning by Interview," the engineer was completely dependent on the availability and desire of her peers to share what they know. In addition, whatever implicit bias her peers may have would likely come out during those interviews. In contrast, in our second example, "Learning by Reading," the responsibility is placed on the engineer to control her career, and more importantly, she is working within a system that allows her to control her career.

As seen in our Causal Diagram (Figure 6.2), Control over Career is driven by Personal Growth which is driven by Control of Learning which is enabled by access to Reusable Knowledge. Access to Reusable Knowledge feeds Confidence, which then drives Personal Development, then Technical Depth, then Business Contribution, and finally leads to Satisfying Work. Her newfound Technical Depth may also lead to more Balanced Work.

Specifically, looking at our causal diagram (Figure 6.2), the learning path that the engineer followed in the "Learning by Reading" story is not by accident, our engineering system has been designed to force it to happen. The engineer's manager did not say something to the effect of, "Well, they keep all their knowledge on a website, but to really understand the work, you should just talk with either the engineers or the manager." Her manager took the opportunity to provide a specific level of coaching. The manager explicitly communicated that it was the engineer's responsibility to learn what she could from the readily available information. Further, the manager informed the engineer that she needed to make printed copies of the most relevant A3 sheets so that when she meets with the team's manager, she can ask specific questions about the work. The engineer's manager knows that the discussion that the engineer will have with the team manager, using the A3 sheets, will give that manager the opportunity to provide a higher level of coaching than her manager could. Any of us in a job search knows that before you go into an interview, you research whatever information you can find in order to prepare for the interview. Managing the technical aspects of a career is no different; however, the system needs to be designed to enable it.

If you re-read Alissa's Medical Intensive Care Unit (MICU) rotation story of creating satisfying work while building confidence from Chapter 4, you will see that this exact flow described in the box above occurred at our kitchen table. Her residency program has been designed to enable it. However, it was then her decision and responsibility to use it. It began with control over career, focused on using reusable knowledge, worked toward building confidence, and ultimately ended with creating satisfying work.

Now that we have developed our engineering system, we will now shift to the specific area of personal or individual development plans.

PERSONAL DEVELOPMENT PLANS, DRIVEN BY THE CAUSAL DIAGRAM

There are dozens of books, websites, and corporate processes used by individuals and management teams that describe ways to create personal development plans. In the context of retaining female engineers, we won't even attempt to provide insight into what makes one better than the other. In addition, we also know that the format and approach change frequently and are usually linked to a performance review that ultimately drives salaries, bonuses, and promotions. So, this may be a place to utilize reengineering – basically, start over.

While there are many different approaches to development plans, for an organization that is focusing on retaining women engineers, we recommend creating a development plan template that is based on both the needs of the organization and a causal diagram that is specific to the current state of the organization. The development plan template should be constructed with the understanding that reusable knowledge is the cornerstone for the success of all. In the development plan template, explicitly identify the current state of each of the critical elements (i.e., the root causes or areas of opportunity) and then identify the areas that are in need of work. The critical elements and the areas in need of work will be specific to the organization. For example, in a progressive firm that has spent a decade or more working on most of the elements related to climate (Figure 6.2), these elements may be in an "ok state." We know that implicit bias always exists and influences each of these climate areas, so we are not suggesting that they not be part of the work, but they may be approached as secondary area of effort or improved by focusing on another critical element. With the elements related to work climate in an "ok state," the firm may identify other areas of focus on their development plan template, like Role Models. At the end of this process, there is a very simple test for the template: if you showed the organization's causal diagram and the template to someone, how obvious would it be that they are linked?

When each individual creates their own development plan based on the development plan template, they should start with their desired work or career direction that is consistent with the goals of the organization and is within the individual's control. While the specific areas of focus will vary depending on the organization and the individual, we suggest that in combination with the organizational efforts to build Reusable Knowledge, the individual development plans for a woman engineer or manager could focus on the three critical areas: Role Models, Balanced Work, and Technical Depth.

1. *Women Role Models*: This area has two components. The first is to create female role models in order to improve the work climate (Figure 6.2). Women role models also help build a diverse leadership team. Improving the development of women role models is particularly important when working with first or second-level women managers* and technical leads.

* In our use of first- or second-level manager, we used the structure of engineers reporting to first-level managers.

In that case, those women identified as role models should have a portion of their development plan highlighting the role model component. The plan should articulate specific actions to promote or grow their role in the organization – no different from any other leadership attribute. This could include an overt discussion with a woman about the need to demonstrate the behaviors and skills of a role model. Although this might sound like yet another added responsibility to women, it is no different from asking her to lead the direction of a business process – it is just part of the responsibility of being a leader. The key difference between this task and other relational tasks is that it is an explicit and valued part of her development plan and therefore contributes to her bonus and promotion potential. This is in direct contrast to the undervalued and underappreciated relational tasks that are frequently asked of women which do not always result in increased opportunity, monetary gains, or career advancement.

The second component is having individuals identify role models who they can emulate. Having role models can help them develop overall career objectives as well as personal development goals, both of which increase their control over their careers. An individual's development plan should identify who they feel are their role models, the attributes of that role model that are important to them, and how they may achieve those specific attributes. The results of these discussions could then be passed on to the specific role models within the organization. Further, having a discussion with men about women role models further helps the effort as these discussions find allies, promote change, and ensure visibility of the organization's effort in this area.

2. *Balanced Work Leading to Satisfying Work*: This area will ultimately be driven by the specific responsibilities of the group. When creating individual development plans with an emphasis on balanced and satisfying work, there should be time intentionally spent looking for ways to broaden an individual's growth. A portion of assigned work should add to the development of the individual. Balanced work could be established by having both system-level and detailed design work, or both technical and customer-centric work within a personal development plan. In addition, the plan could be as explicit as setting a goal to improve an existing development activity such that it takes half the time to complete. The idea of balanced work, or in some cases breadth verse depth, can enable "T-Shaped Skills." The idea behind "T-Shaped Skills" is that the vertical bar represents technical depth and the horizontal bar represents a collaboration with others and the application of knowledge outside their own expertise. This is where it may be required to address the gender role issues (Relational Practices) that women may have naturally moved into. Additionally, there may be a need to "rebuild" the vertical bar of the "T" – technical depth. The discussion around balanced work should include developing an understanding of where bias might play a role in selecting that work. For example, it may be

those specific roles, although technical in nature such as product test development, have been classically led by women. So, it may be time to enable women to branch out away from this type of role.

3. *Technical Depth*: This is the area that should recognize that enabling managers and women to remain technical or build technical capability is critical. As we saw in Chapter 5, "The Career Path" section,

   > For women, being on the technical path was favorable to the(m) being on the managerial path in terms of intent to leave engineering; the hybrid and technical paths were favorable to the managerial path in terms of identification with other engineers and meaningful work
   >
   > **[1, p. 105]**

   So, while we encourage those on the technical path to focus on technical skills as part of their development, we want to do the same for managers. As we show on the causal diagram (Figure 6.2), technical depth is linked to both satisfying work and business contribution. For managers, technical depth is built through personal development efforts that enable a contribution to both the work of (product) development and the coaching of engineers.

As we work through these three key areas as part of the development plan, we want to note how they relate to three other areas: relational work, control of learning, and control over career

Relational Work

In Chapter 10, we used the story of the experienced male engineer handing his work off to the woman program engineer for validation and implementation (see: An Engineers Responsibility: Detail Design through Implementation). From an engineering system design perspective, this story addressed a number of issues, including eliminating a handoff that contributes to either creating or covering up waste – which could be measured in the amount of time to complete the work. The basic purpose of relational work is to bring people together or integrate work across multiple groups. When we look at relational work, we do not want to assume that the right thing to do is to just spread it out across multiple people. We want to start with the fundamentals of lean, which is eliminating waste. To understand how to approach the management of relational work, we need to start by understanding the three types of work that exist within a process and how they might fit with relational work.

- Value Add – An example could be group meetings that drive innovation or eliminate waste. These efforts have measurable gains that the organization sees. We can assume that their highest value add comes through reviewing A3s as part of doing the work.
- Non-Value Add – For example, non-technical meetings that seem to make people feel good that things are being coordinated but could be handled

and managed with well-done Program A3s that are reviewed in small group settings as needed.
- Non-Value Add but Necessary – This is the relational work that is specifically looking for disconnects within the work across teams and establishing the actions to address them. This might be managed by having a weekly review of a program level of A3 with critical individuals to test with members and identify problems and establish the required action. While this does not directly result in increased value to the customer, they are required for the overall function of the organization.

The purpose of examining relational work, and specifically how it has been delegated is not to suggest that none of it is needed but to look at who is responsible for it, how it is managed, and how to maximize its value.

Control of Learning

The starting point for this area is through the A3 reports that an individual is responsible for. Whether these are A3 reports an individual directly owns through the activity of doing her/his job, or, as a manager, the ones that her team is responsible for and she is regularly reviewing or contributing to as part of her leadership responsibilities. In both cases, A3 reports play a central role in the management of a career. Each time an individual shares their A3 reports with an individual, in a small group, or as part of a review process, there is an opportunity to learn or to identify areas for future learning.

To promote structured learning, a development plan should identify specific areas of focus that are a part of the group's work or some aspect of adjacent work that supports overall career plans. This area should be treated like earning a college degree, with defined areas of learning and expected progress. As the organization begins to build its level of available knowledge, an individual can use that knowledge to work toward specific focus areas. In some cases, it may be the growth of the horizontal bar of the T-Shaped Skills or adding another vertical bar to create what is now being promoted as "Pi-Shaped Skills" (using the Greek Pi Symbol, π).

Control over Career

If the above areas are being worked on, then control over career should be possible. This is the objective of a personal development plan. As an individual, the question is: Do I have things put in place that allows me to control my career? The fundamental test is whether someone can execute learning by reading. As engineering students, the most critical pieces of information or knowledge that we needed in order to obtain our desired engineering specialty were: the piece of paper that listed the course requirements by term, the course syllabus, and being able to buy the textbook for the class. The person who taught the class may be important, but not critical. Similarly, each individual has to have the organizational structure in place to build a comparable map for their career.

THE CONTRIBUTION OF MEN VIA THE CAUSAL DIAGRAM

Reviewing the organization's causal diagram with male peers (engineers and managers) to understand what part of their development plans might directly address an area of opportunity works toward the broader objective. This is where the organization's goals and focus come front and center. The discussion about a contribution should go beyond what male peers may be doing as part of their work in the creation of available knowledge or enabling Lean Development within the organization. Their development plan should proactively pull them into the efforts to contribute to broader solutions including retention and professional development of female engineers, like supporting female role models.

ONE-ON-ONE REVIEWS OF A3 REPORTS – DEVELOPING THE ENGINEER

In Chapter 11, we worked through the use of A3 reports as a learning tool and in Chapter 13 we explored the use of A3s from an organizational perspective. Expanding on those purposes, A3 reports can also be used as a tool for developing an engineer or manager. In our use, the development of an individual goes beyond coaching. Coaching may largely focus on building skills and capability in the direction needed to solve a current problem. Personal Development is specifically looking for other directions to point the engineer or manager toward in the context of her broader career goals. As mentioned earlier, this might be a specific area of new technical knowledge or skills that would benefit someone or an aspect of building a new area of customer or business understanding.

In Chapter 10, our story of "Coaching Set-Based-Design" and the use of an A3 report were focused on learning how to solve the current problem. In that context, we can overtly use the A3 process to focus on learning how to solve the next problem. This means that we are looking for opportunities to teach an engineer "how to learn" and "how to think critically," rather than just trying to improve their ability to get the work done. An example might be one in which the current problem needs a specific task completed that the group's resident expert might normally do. From a personal development perspective, the engineer might be asked to have the resident expert teach her how to do that task through the use of an A3. The engineer would then complete it on her own. This activity would never have been known during the creation of a personal development plan, but through the A3 process of solving a current problem the opportunity becomes visible. That visibility allows for increased growth of an individual.

None of this is extra work. It is all part of the same work of solving the problem and is something that occurs in real time. This opportunity exists every time there is a review of an A3 report. In addition, it is not part of some year-end planning function; rather, it is managed as part of doing the job throughout the year. In its most basic form, this process could begin at the very first meeting held to talk about the problem and the need to create an A3 report. The manager may ask the engineer whether they know what a causal diagram is and how to use it. Through the conversation about the causal diagram, the discussion moves to discussing the trade-off curves and the manager asking, "How would you create the learning to draw it?" This process could also occur through longer-term discussions, such as the need to have someone pick up responsibility for a new development area.

The A3 review process that may be occurring weekly is centered on accelerating the resolution of the problem at hand while looking to the future for other opportunities, which ultimately creates satisfying work.

STRUCTURED LEARNING – ADDRESSING THE CONFIDENCE GAP

The gender confidence gap is unique to each individual. Based on prior research and publications, on average, the gap is real. As with all areas of our gender schema, this indicates a typical tendency or expectation, not what is necessarily true for a particular individual. Whether, for a particular individual, we are discussing confidence as the internal measure of their confidence or an observer's perception of a woman's confidence, which is shaped by how she is expected to behave in professional environments, is not the most critical thing to establish when we are focusing on an individual's development plan. The key aspect in this area is what matters to the individual engineer. In working to mitigate the influence of gender confidence, we have focused on two aspects:

1. Providing available knowledge to all, noting the specific benefit women see from this
2. The use of A3 reports to manage the development work and provide a tool to address both internal confidence and the perception of confidence by others (which may just be bias)

A3 reports are a tool that individuals can use to build their confidence. The information communicated in written form can minimize the bias that might exist if that same concept was verbally communicated.

In establishing the development activities for an individual with an eye toward building confidence, the first thing to understand is whether that area needs focused work. If the answer is no, then great, move on. However, if focused effort is needed, then a plan should be put in place to address it. Based on the work described in the previous chapters, we might break the plan into four areas:

1. Understanding where lack of confidence might exist
2. Confidence in the content presented on A3 reports
3. Confidence in presenting A3 reports or design work to larger groups or challenging one-on-one settings
4. Broadening a skill set to adjacent areas

As an example, for a knowledge review led by a new-to-career individual that requires "spinning CAD" and likely includes senior-level designers, the manager should be clear about the need to have the engineer fully prepared to lead the CAD review, which means taking their best mouse and appropriate computer into the room. Getting to the room early to set up. Pulling up the CAD files necessary for the reviews and making sure the CAD tools function as expected. This is not the time to show up right on time and then struggle to get things set up. For the male engineer, those frustrating issues that arise in a review may be viewed as just the norm and may even be funny to the group. In contrast, given our gender schema and unconscious beliefs about competence, for the lone women in the group, these issues will likely not be viewed as favorably.

CONCLUSION

Many organizations across various industries have a vast variety of methods for enabling individuals to create personal development plans that support their professional goals. For those companies that are generic in their development plan templates, they may not be effectively capitalizing on the use of personal development plans to address broader objectives for the firm. In this chapter, for the progressive firm that is focused on retaining women engineers and managers, we are suggesting that a personal development plan starts with a template that is based on an organization's needs identified from an organization-specific causal diagram. Each individual starts with the firm's overall efforts to create reusable knowledge, move the work toward the A3 Problem-Solving Process, and use Set-Based Design and System Thinking. They then identify their own specific areas of need. A well-created personal development plan that is driven by the organization's causal diagram creates the opportunity to establish specific individual learning and personal development opportunities while simultaneously improving retention of women in engineering.

Through the creation of strong personal development plans, we have moved the organizational change from the highest leadership level all the way to how the work is actually done and how to meet an individual's professional goals.

REFERENCE

1. M. T. Cardador, and P. L. Hill, "Career Paths in Engineering Firms: Gendered Patterns and Implications," *Journal of Career Assessment*, vol. 26, no. 1, pp. 95–110, Feb. 2018, doi: 10.1177/1069072716679987.

15 Leading Beyond the Causal Diagram

We started this journey with the current situation despite two decades of work:

- In 2019, women made up only 22% of engineer graduates with bachelor's degrees [1]
- Women make up only 15% of engineers in the workforce [2]
- Meanwhile, women entering medical school have grown beyond 50% [3]
- Diversity drives better business performance and customer solutions [4]

In Part II of this book, we articulated a myriad of factors that deter women from choosing engineering as a career and, if they do choose it, those same issues may derail their goal to make it a lifelong career. Many of those factors are important to some women, but not others. Further, the influence of those factors may be driven by stereotypical beliefs and expectations about women in the workplace due to our underlying gender schema. Ultimately, when looking at individual women considering engineering or already in the field of engineering, the factors and their effects are just averages. Although moving the average of anything up helps, it does not work when you are dealing with systemic issues. Based on our causal diagram (Figure 6.2) and our knowledge of how physicians are trained, we believe that the fundamental problems in the system of engineering are twofold: how engineers are trained and learn on the job, and how they do the work. At the center of these issues is a lack of access to knowledge. We have covered specific areas regarding how to change the work of engineering and how to change the learning of an engineer but, before we summarize those, we will start with some thoughts on approaching the efforts from a management perspective.

Our change efforts need to be grounded in changing how the work and learning are done. In addition, we need a broad, intentional movement that begins at the top. So, we suggest the following:

1. Make the direction of the change clear, concise, and regularly reinforced throughout the organization through how the work is done.
2. Work toward creating an organization in which the work environment is intentionally structured to be a learning environment.
3. Work toward an organization in which the structure of the work is designed to minimize the impact of bias and bring out the best work in everyone.
4. Make a thoughtful choice as to whether a woman or man should lead the change at the various levels of work. Male leaders act as allies; however, the

DOI: 10.1201/9781003205814-25

consideration of how much political capital* an individual woman has in order to lead these efforts and what risks she is willing to take needs to be understood.

5. While this type of change may be viewed solely as a lean transformation, in reality, it should be approached as a vision of what an organization (and ultimately the profession of engineering) can become.

Throughout the course of the book, we have used the causal diagram, Figure 6.2, to guide our recommendations to change the way the work is done and learning occurs, all through the methods of Lean Development.

We assert that training in medicine can be a model for training in engineering (Chapter 4). At a basic level, the difference between how a physician approaches learning and problem-solving compared to many engineers is driven by access to a vast amount of knowledge. Equal access to this knowledge creates a more level playing field for a female physician through control over her learning, increased confidence, and, ultimately, the ability to engage in satisfying work and having control over her career.

Our causal diagram (Figure 6.2) is intended to show the link between low retention rates of women in engineering and the lack of reusable knowledge being a root cause. This lack of reusable knowledge can limit an organization's ability to create role models, negatively affect control of learning, hamper the ability to build technical depth, and hinder confidence which can increase bias (see Figure 2.1).

In addition to the importance of reusable knowledge, the second key piece is to change the method of work and understand the challenges of Build–Test–Fix (Figure 9.1). The Build–Test–Fix approach creates an environment in which an engineer's confidence in her selected "best idea" and experiences becomes a leading factor in the delivery of a design solution. Build–Test–Fix ultimately expects the first attempt to fail and plans for it by creating project milestones specifically with repeated, and hopefully improved, attempts. Our promotion of Set-Based Design (Figure 11.1) puts the identification of multiple options as the central focus, followed by the use of quick learning cycles to identify and eliminate the weakest solutions. This method yields the strongest solution in the end, delivers a better product faster, and creates a climate in which the team or individual is focused on learning. It promotes a learning culture and moves the discussion from opinion, driven by experiences, to facts, driven by knowledge. As engineers work through the learning process, they naturally build their knowledge in a new development area while working toward a design that delivers the highest customer value. Finally, we have identified the need to ensure all engineers are coached in a structured manner that promotes systems thinking and further expands the engineers' knowledge base, preparing them to solve the "next problem."

We have promoted the use of the A3 Problem-Solving tool as a method that engineers and the management team can use to deliver a new product, process, or service.

* Political capital in this context is specific to how far a woman leader can move beyond the organization's expected behaviors of her while not jeopardizing her current role or future career objectives.

This tool changes the way the work is done. Now, let's use that same tool, an A3, to develop a plan for an organization's effort to implement Lean Development with two purposes:

1. Attract and retain women engineers and managers
2. Deliver better business results through improved efficiency and the creation of a diverse work environment

An organization can elect to change the order of these because the same work results in both benefits. However, in this order, it puts our overall objective at a broader level than just profit. By focusing on attracting and retaining women, the management team is targeting issues that affect half of the people (women) an organization would hope to attract or retain – with the added purpose of maximizing profit through increased diversity.

In the creation of this A3, there are a few key points:

1. Be clear and concise about the goal and current situation (the left side).
2. The management team needs to create its own organization-specific causal diagram, or some other clear list, to identify where it has issues and where it has strengths. This causal diagram or list communicates the opportunities for growth and therefore structures the work moving forward.
3. From the organization-specific causal diagram or list, identify the critical issues. Initial efforts need to focus on these critical issues, using an approach grounded in changing the way the work is done and how learning occurs.
4. Establish how the work flows through the organization. This might begin with where business opportunities are first identified. Outlining the workflow shows where there are development opportunities for women. These opportunities may be central to creating satisfying work and enabling control over career.
5. Assume the A3 will structure and manage a year's worth of work.

CREATING AN INCLUSIVE WORK ENVIRONMENT

Throughout this book, we have asserted that a progressive firm may have already implemented many aspects to enable an inclusive work environment by targeting elements such as work–life balance, being valued, being respected, and accepting diversity. However, the organization needs to look at how much more work needs to be done in this area and how the work is done. It needs to assess if the basic foundation is in place. The progressive firm needs to establish whether having one more presentation on Implicit Bias to a group of white males, who have probably already heard it and have an understanding of it, is needed. A tool to help those involved align their understanding of the state of the current work climate and identify the direction needed for work moving forward might be to use the trade-off curve in Figure 9.2. For each area of climate (along the diagonal progressive lines), each woman could score her experience from 1 to 5 in order for a group to create the trade-off curve.

CREATING REUSABLE KNOWLEDGE – ATTACK THE GENDER CONFIDENCE GAP AND GENDER BIAS

In Chapters 13 and 14, we wrote at length about making the A3 process central to how the work is done. We spoke of how it is fundamentally a tool that enables both top-down and bottom-up change. It is a tool to use for coaching, delivers the highest customer value, brings out the best work in people, and eliminates opinions so that we can focus on clarity, conciseness, facts, and the unknown. The action here is to start using them and putting them at the center of the work. It is through this tool that we can positively influence confidence and therefore reduce bias in the long-run and ultimately improve the work climate (Figure 2.1).

In moving to the use of A3s, the organization needs to find a simple approach for ensuring that everyone in the organization has visibility to the A3s that are created. It is best to start simple, with a current file-sharing tool and a clean area to store the new knowledge, with an expectation that things will need to be changed in the future. The broader visibility of this knowledge will enable individuals to take control over their learning.

CREATING A NEW WORK AND LEARNING ENVIRONMENT

Through the use of A3s in the development work, a focus on systems thinking, the use of Set-Based Design (Figure 11.1), and the use of coaching via an engineer's (or manager's) A3s, an organization can build a structured approach to both do the work and enable learning. This learning process starts first with the business objective that the organization is trying to accomplish, and then focuses on the flow of the work required to accomplish it (Figure 13.1). The engineer working through the development work or knowledge creation has the opportunity to see firsthand how her work directly contributes to broader level business goals and how those goals flow down to the work she is doing. She has the opportunity to ask questions about the higher-level plans and improve her contributions. The creation of her A3s and the coaching she receives during reviews naturally expands her knowledge of the development area as she works through solving the problem or delivering the solution. The use of Set-Based Design requires her to think through her own areas of required learning and how she will measure the success, or failure, of any solution she considers. A discussion about her understanding of the design space or the technical relationship between design alternatives through the creation of any needed trade-off curves (Figure 11.2) naturally strengthens her confidence in the work she is responsible for and ultimately delivers a better solution. Each time someone sits down to have a technical review of her work, via her A3s, she has the opportunity to learn, similar to her work in obtaining her engineering degree.

CREATING ROLE MODELS

Adding to or creating women role models best occurs when there is a management objective to do it. The creation and visibility of women role models and their contributions will improve the organization's work climate and deliver better business results. Within a progressive firm, which is maybe 20% women (engineers and managers), women role models may exist; however, the leadership team can establish a clear plan to increase the visibility of their contributions and grow the numbers. Access to reusable knowledge allows a role model to expand their knowledge base. Further, their contributions in the review process of A3s created by less senior engineers will naturally maximize their impact on the organization and its people.

In the Chapter 14 story, "Learning by Reading," we highlighted how a new-to-career engineer was able to add a degree of control over her career by having access to the reusable knowledge created by another team in an area of work that she was interested in. She then worked with the manager of that team to understand specific details through the A3s she thought were most pertinent to her. With that example, let's now play out how that experience is expanded to promote the role of an up-and-coming woman role model, specifically a technical lead.

CREATING ROLE MODELS – FROM LEARNING BY READING

We begin with the manager telling the engineer to approach her learning using the other group's knowledge repository and A3s. However, in this story, the manager suggests that instead of talking to the group's manager, the engineer should talk to a newly promoted female technical lead within the department about what the engineer might want to accomplish. After the engineer has reviewed the team's knowledge sheets as well as identified and printed the few sheets that will guide the discussion, the engineer sets up a time to talk with the technical lead. The technical lead has experience in an adjacent development area but is not an expert in this specific area. The lead accepts the meeting invite and, in preparation for this meeting, she spends some time looking at the other group's knowledge repository.

Why did she do this? Because her manager has had explicit discussions with her about an organizational objective to develop women role models. In fact, becoming a role model is part of her job as a technical lead and being prepared for this type of discussion supports that effort. With her own contributions in creating reusable knowledge for her work, which she is an expert in, she is well-versed in how the other team would have approached creating their reusable knowledge. In addition, her review of their knowledge naturally expands her areas of technical expertise and she is looking forward to learning about this area as much as she is looking forward to the coaching experience.

When the engineer and technical lead meet to review the printed A3s that the engineer has brought, the technical lead provides input about the A3s and adds additional perspectives to the content of them given her technical knowledge. As they talk about the sheets in front of them, the lead pulls up the group's knowledge repository. From the team's site, the technical lead is able to identify other sheets that fill in additional technical questions. The lead may find a trade-off curve that directly relates to the engineer's own development responsibility. The lead may look for an A3 that demonstrates the team's use of Set-Based Design. Using these, the lead is able to explain to the engineer how when people have access to detail technical knowledge, they can integrate across multiple technical areas and deliver a better solution.

The engineer and technical lead end the meeting in agreement that the engineer should now sit down with the engineer's manager and see how this new technical knowledge might be worked into her current job. Based on this discussion, the manager and engineer can also talk about whether this might be a development area that she wants to move into.

Finally, because both engineers prepared for this and used A3s as the basis of their discussion, this might have been a 45-minute meeting. Additionally, this entire effort sits on the 45-degree line of our Chapter 10 "Knowledge versus Time" graph (Exercise 10.1).

For any organization trying to build women role models, the above story should give an example of how it might work. However, without some defined structure that is part of how the work is done, the results are unpredictable. Imagine if we played out the Chapter 14 story of "Learning by Interviewing" in which the engineer's manager directed the engineer to meet with the female technical lead to talk about a technical area she does not work in – as opposed to the design engineers in the group. In this situation, the engineer might see no bias and have a much more positive experience than interviewing members of the technical team. However, without the female technical lead being an expert in that area or having the ability to gain knowledge in that area through access to reusable knowledge (i.e., without access to A3s), the female engineer would receive less detailed coaching and would not see the true capability of the technical lead. Similarly, the technical lead would not see herself as an effective role model or mentor and therefore not build additional confidence in herself.

In an engineering system centered on the principle of Lean Development, both of the individuals in the story mentioned above were able to maximize their personal development opportunities.

OTHER AREAS OF FOCUS

Below are two areas of consideration related to the social aspects of an organization that is beyond the usual areas of focus of Lean Development. These areas are proposed for further consideration by a firm's leadership.

Finding Opportunity and Networking through A3s

Networking consists of people connecting through the sharing of information. In most cases, that information might be focused on career opportunities. In the case of the use of A3s as a critical piece of the development process, those same A3s become a natural vehicle to share information which can then build a person's network. The information on the A3s may identify someone's technical, business, or leadership experience and responsibilities. This experience could range from detailed technical knowledge to a new business opportunity that someone evaluated and others are now able to see. This sharing of information facilitates a connection between individuals within the organization, just like we saw between the technical lead and the engineer in our story above.

Creating and placing A3s in a well-managed digital repository allow individuals to find people with knowledge in areas they want to learn from or align common goals. Reusable knowledge, therefore, can help someone find opportunities that increase their level of control over their career (Figure 6.2).

The Purpose of Meetings – Problem-Solving or Communication

An internet search for "purpose of meetings," "effective meetings," or "efficient meetings" results in dozens of suggested formats, goals, and expectations. However, when we think about the purpose of a meeting, as engineers, we naturally break it down into either problem-solving or communication (within problem-solving we would include planning or decision-making). Durward Sobek II compared these two forms of meetings between Toyota and Chrysler [5]. In that review, he found that Toyota would "Focus cross-functional meetings on solving a specific problem(s); leave background communication to written media and informal channels" [5, p. 174]. Additionally, they worked to "Minimize meeting time through judicious use of written communication; meet on an ad hoc basis" [5, p. 174]. However, in looking at Chrysler, Durward writes,

> They must emphasize communication, because Chrysler has few systems in place outside of centralized computer data bases for high-bandwidth communication. As a consequence, Chrysler's process is characterized by regular, weekly or daily meetings such as Spaghetti Day,* pilot build, and departmental design review meetings.
>
> **[5, p. 174]**

Thus, Chrysler must "Use cross-functional meetings for communication as well as problem-solving" [5, p. 175]. Furthermore, Chrysler "Organize(s) the product development process around regularly scheduled cross-functional meetings" [5, p. 175].

So, with this research being published in 1997 and there being a high likelihood that Chrysler has evolved their meeting methods since that time, why does this matter? It matters because this comparison provides us with another view of an organization's development process and the interaction amongst people within those meetings. In particular, when we look at an organization's use of meetings for communication†, we need to understand how those meetings can increase the effects of

* Spaghetti Day, weekly all day meetings to review development status [5, p 109].

† We assert that meetings that are used for problem-solving are managed via the A3 problem-solving process, whereas meetings used for communication may be using slide decks or just general discussion.

bias or, at the very least, do nothing to mitigate bias. If we recall the biology class study that we described in Chapter 3, Eddy et al. found that male students spoke roughly 60% of the time even though they were only 40% of the class [6]. In a standard communication meeting, therefore, we would expect that these same dynamics would occur and be magnified as men outnumber women in engineering. Further, we know from our discussion of gender schema in Chapters 2 and 3 that women are perceived to be less competent than men and that there are specific expectations for how women should act in the workplace. Communication meetings, therefore, run the risk of increasing the negative effects of bias, which can then have significant downstream effects. In contrast, meetings used for problem-solving can mitigate the effects of bias as they are structured around an A3, with the knowledge and understanding of the engineer clearly written down and the discussion in the meeting is based on facts and knowledge rather than opinions.

With an overall objective of reducing bias, the use of communication meetings, which would include weekly team meetings or unstructured daily standups, needs to then be evaluated through the lens of the personal dynamics that can occur within them. The key is to look at communication meetings from both an organizational need (how do they add value and how do they promote Lean Development) and whether or not they promote bias within their current method of use. This may also be a time to follow the lead of Toyota from 25 years ago: "Minimize meeting time through judicious use of written communication; meet on an ad hoc basis" [5, p. 174].

PROGRESS FORWARD

If we look back at our last story of the new-to-career engineer working with the female technical lead, who was in the process of becoming a role model, we can see the larger impact of individual events. In this case, we can see that this one event, an engineer wanting to understand a new design area and how it was approached, has driven improvement in almost all of the 14 "opportunity" bubbles on our causal diagram – 12 of them to be specific. As we have stated in most of the previous stories, seeing this kind of improvement is not by accident, it is by design, specifically the intentional design of the engineering system. Through the use of the causal diagram and Lean Development methods, we have designed a system that will move the organization forward.

We use the words, "Progress Forward," to reflect the idea that without forward movement you are falling behind. The focus on improving the work climate in order to facilitate a change that increases the success of women in engineering has been present for decades. The efforts to attract girls to Science, Technology, Engineering, and Math (STEM) fields have delivered good results in almost every area except engineering. For this reason, we assert that although there have been significant efforts over the last two decades and the methods and actions taken to date are important, the results show they are not sufficient. Something is wrong; some things need to change.

For the progressive firm that has focused on creating an inclusive work environment, the fundamental issue impacting the growth of women in engineering is "how the work is done" and "how learning occurs." Through the methods of Lean Development, a firm can improve its business competitiveness while at the same time helping to retain women within its organization and the field of engineering.

In working through our goal to improve the retention rate of female engineers, we have actually changed the work environment through the use of Lean Development. Bob once overheard a discussion during a mock interview in which a future computer science graduate was speaking to an individual acting as a potential employer. The future graduate asked a very simple question, "do you use Agile* within your organization?" Bob got the impression that this candidate may not seek employment with a company that does not use this method. The candidate may even "run for the hills" if the interviewer replies with, "No, we tried that once, but our culture just doesn't work like that." As development organizations adopt the methods of Lean Development and the use of A3s, and university engineering programs teach these methods, the female engineering candidate may someday be asking "Does your organization use the principles of Lean Development and specifically do they use A3s?" For the female engineering candidate, this may be the deal breaker between the multitude of options that she has.

CONCLUSION

The causal diagram (Figure 6.2) provides the technical overview of the factors contributing to our problem. Some of these factors are root causes and others are just critical. However, the causal diagram does not give us the level of strategic insight to implement the broad level of change needed for the problem we are dealing with. For that, we turned to the six principles of Lean Development, the implementation of which can provide the necessary system-level change. In this chapter, we summarized six areas of organizational-level effort that can enable that broad level of change and address the underlying root causes and critical factors. Each of the areas either focuses on the mechanics of the work, such as A3s to create reusable knowledge, or the social needs of an organization, such as an inclusive work environment or the development of role models. Each of them, and you may find more, shift the focus of the work from completing a tactical set of actions to working toward a broader vision of what an organization or the field of engineering might become. We hope that the "gray box" stories throughout the book have demonstrated both what the current environment may look like and what the future environment might become. Finally, we hope that the methods proposed in this book, to change the system of engineering through the implementation of Lean Development, enable every woman to make engineering a lifelong career and improve the overall work of engineering for everyone.

---

* Agile methodology is a type of project management process, mainly used for software development, where demands and solutions evolve through the collaborative effort of self-organizing and cross-functional teams and their customers [7].

REFERENCES

1. American Society for Engineering Education, "Engineering and Engineering Technology by the Numbers 2019," American Society for Engineering Education, Washington, DC, Report, 2020. Accessed: Dec. 03, 2021. [Online]. Available: https://ira.asee.org/wp-content/uploads/2021/06/Engineering-by-the-Numbers-2019-JUNE-2021.pdf.
2. Society of Women Engineers, "SWE Research Fast Facts," Society of Women Engineers (SWE), Sep. 2021. [Online]. Available: https://swe.org/wp-content/uploads/2021/10/SWE-Fast-Facts_Oct-2021.pdf.
3. AAMC, "2021 FACTS: Applicants and Matriculants Data," AAMC, 2021. https://www.aamc.org/data-reports/students-residents/interactive-data/2021-facts-applicants-and-matriculants-data (accessed Dec. 02, 2021).
4. V. Hunt, L. Yee, S. Prince, and S. Dixon-Fyle, "Delivering Through Diversity," McKinsey & Company, Report, Jan. 2018. Accessed: Jan. 15, 2022. [Online]. Available: https://www.mckinsey.com/business-functions/people-and-organizational-performance/our-insights/delivering-through-diversity,
5. D. K. Sobek II, *Principles that Shape Product Development Systems: A Toyota-Chrysler Comparison*. Ann Arbor, MI: University of Michigan, 1997.
6. S. L. Eddy, S. E. Brownell, and M. P. Wenderoth, "Gender Gaps in Achievement and Participation in Multiple Introductory Biology Classrooms," *LSE*, vol. 13, no. 3, pp. 478–492, Sep. 2014, doi: 10.1187/cbe.13-10-0204.
7. Agile Alliance, "Agile 101," *Agile Alliance*, 2022. https://www.agilealliance.org/agile101/ (accessed Jan. 29, 2022).

Part V

Summary

In Part V, we described three elements to facilitate overall work satisfaction.

- In Chapter 13, we focused on the structural changes that move an organization toward the use of Lean Development and described how those changes are implemented within the organizational structure. We discussed how the use of Lean Development can enable work–life balance and therefore create a more positive work environment.
- In Chapter 14, we described a key element of lean, the development of people. We focused on how deliberate coaching through the use of the Lean Development principles gives a leader a playbook for doing that work. In our hypothetical story of "learning by reading," the manager provided direct input to the engineer on how to look at the A3 relative to her own work, but more importantly taught her the value of this method through tangible actions.
- Finally, in Chapter 15, we provide specific recommendations on how to approach changing the work of engineering from a management perspective. Two of the recommendations in particular are unique to our effort. The first is to work toward a work environment that is intentionally created to be a learning environment. The second is that although the change can be approached as a lean transformation, there is an opportunity to create a vision of what an organization can become and to use the change to work toward that vision. In order to work toward this broader vision, we identified different areas of focus that draw upon the principles of Lean Development and help us move beyond the causal diagram.

DOI: 10.1201/9781003205814-26

Afterword

*Getting Back to Finding the Longitude**

For centuries, the scientific community believed that the only way to navigate the world's oceans was from the stars and moon. Within these beliefs and debates were ideas that were based on scientific knowledge as well as those that were personally motivated. For those who did accept the possibility of other solutions, there was still a lack of belief in man's ability to design and build a tool that could survive ocean travel for weeks while delivering the required accuracy. Without knowing the longitude and without having a way to determine it on a ship's rolling deck under weeks of cloud-covered night sky, the ship's captain and his sailors were basically lost at sea. For nearly four decades, the inventor of what would become a chronometer, accurate enough to reliably find the longitude after a six-week journey at sea, toiled through the creation of five versions of his clock designs. He spent years making each of them and then years adjusting them, with each design improving on the last. His design began with the requirement that it must be accurate within three seconds a day while able to deal with changing humidity, pressure, and the ship's rolling motion. He created designs so advanced for their time that some are used today. In our case, we are not promoting new, experimental change that is at the forefront of technology or organizational understanding. The changes we are promoting are now over 20 years old with more than a dozen years of research behind them. Similarly, the challenges that women face in STEM have been communicated for decades as well. The question then becomes, what action is needed, what action is to be taken, and most importantly, what is preventing us from starting?

As we look at the current issue of low retention of women engineers that we outlined in Chapter 1, we can evaluate how the work is done, what tools and methods are being used, and what opportunities present themselves. We hope that the methods and thoughts of this book promote the adoption of Lean Development to improve the retention of women in engineering or, at the very least, create a dialogue about the need for change and how best to approach it.

REFERENCE

1. D. Sobel, *Longitude: The True Story of a Lone Genius Who Solved the Greatest Scientific Problem of his Time.* New York: Penguin Books, 1996.

* The historical context for this section was drawn from Dana Sobel's book *Longitude – The True Story of Lone Genius Who Solved the Greatest Scientific Problem of His Time*, 1995, Penguin Books [1].

Appendices

Appendix A: Going Home on Time – Lean Development

The Principles

CONTENTS

We have chosen the title of this appendix "Going Home on Time" because unlike many areas of lean, which are focused on improving the process from a customer or business perspective, Lean Development is overtly centered on improving the process from the perspective of the people doing the work. In our case, engineers and developers create new customer solutions. A customer benefits because innovation is done at a faster rate while meeting their needs. The business sees benefits through increased profit and revenue. Ultimately, we can view the principles of Lean Development as a set of personal productivity tools.

PURPOSE OF THIS APPENDIX

The purpose of this appendix is to provide a cohesive overview of Lean Development. This appendix focuses in particular on the six principles that are regularly used to establish the method and thought process of Lean Development. An organization can dramatically improve its competitiveness and overall work environment after recognizing the importance of understanding value as well as seeing that waste is a primary impediment to maximizing value. The principles of Lean Development allow an organization to maximize value and minimize waste.

Although we will avoid directly duplicating content covered in the preceding chapters, we intend to make this appendix a supplement. Therefore, we may summarize content and direct readers back to relevant chapters.

Appendix C contains a list of books focused at Lean Development that can be helpful in deepening your understanding of the methods.

THE BUSINESS CASE FOR LEAN DEVELOPMENT

The initial understanding of the improved efficiency of a development process, that ultimately was termed Lean Development, was published in 1987 in a paper titled "Product Development in the World Auto Industry." In their paper, Clark et al. established that:

- "Japanese projects were completed in two-thirds the time and with one-third the engineering hours of the non-Japanese projects" [1, p. 740].

From a process management perspective, we translate this to being four times more efficient.

Similarly, Japanese automakers saw a competitive advantage compared to US automakers in the area of product quality. In the 1991 book *Product Development Performance*, the authors established a measure of:

- Total Product Quality (TPQ) in which Japanese automakers had 1.7 times advantage over US automakers [2, p. 161].

Therefore, the business case, as established 30 years ago and validated through dozens of books and case studies since then, can be summarized as follows:

> Lean Development is four times more efficient at delivering value to the customer. This in turn means higher profits, higher revenue, and more innovation than traditional development.

Although four times more efficiency was established from research that was conducted 35 years ago, Toyota, as an example, is currently in the number 1 or 2 position among automotive companies from around the world.

A VERY BRIEF HISTORY OF LEAN DEVELOPMENT

Chapter 7 gives detailed information about Toyota's rise to success and summarizes a few of the many books published about why that rise to success occurred. In the area of Lean Development, Toyota's success has been researched and promoted over the last three decades. This was most visible in the publication *The Machine that Changed the World*, published in 1990 [3]. In their chapter "Designing the Car," the authors utilize the term Lean Product Development as well as Lean Design. Further research in the mid-1990s resulted in publications like "Another Look at How Toyota Iterates Product Development" [4] and "Toyota's Principles of Set-Based Concurrent Engineer" [5]. These articles described a "different way" to conduct product development. Sobek's published thesis, "Principles that Shape Product Development Systems: A Toyota – Chrysler Comparison" (1997) [6] conducts a point-by-point comparison of Toyota and Chrysler through the lens of Lean Development and Traditional Development. For example, Sobek describes how Toyota negotiates supplier component specifications through an iterative process in which both Toyota and the supplier align the supplier's proposal with Toyota's subsystem targets through a set-based approach. In contrast, Chrysler's approach is to make changes until objectives are met [6].

THE FUNDAMENTALS OF LEAN – ELIMINATING NON-VALUE-ADD ACTIVITIES – WASTE

As we wrote in Chapter 6, work is comprised of two basic activities:

- Value Add (to the customer) (VA)
- Non-Value Add (NVA)

To broaden our understanding of waste and manage its elimination as a process, we add a third component:

- Non-Value Add but Necessary (NVA but Necessary)

Non-Value Add but Necessary are those activities that we know do not add customer value but we do not see (at this time) a path to eliminating the activity. A simple example is that of dedicated product functional test at the end of a production process. We know just testing the product does not add value; however, it is necessary to find defects or performance issues that we don't want to send to the customer.

In addition, in Chapter 7, we wrote that the typical process that was designed in a typical fashion and has not had lean efforts applied to it (and for the most part is just running) has at least 50% waste in it. So, when we set a goal to improve the process, we should be thinking of a goal of 50% improvement, not 10 or 20%.

The Principles of Lean have been summarized in the book *Lean Thinking* [7] to be:

1. Value
2. The Value Stream
3. Flow
4. Pull
5. Perfection

Through these principles, a process owner or leader effectively has a playbook to maximize the contribution of a process to the business. This playbook allows an organization to move itself to the right on the *x*-axis of Figure 9.2. Further, this playbook ultimately improves the climate. These five principles establish the flow of working toward perfection, beginning with establishing the value. They lay out how people should be approaching the work of eliminating the waste. However, the first step is to recognize that even if the waste cannot be seen at this time, it exists and is therefore an opportunity also exists.

Lean in Manufacturing

As an example, within a manufacturing process, we can look at the flow of the work as the physical movement of materials progressing from one state to the next. In this case, at each step, value may or may not be added. By the time it reaches the final end state, the item has some established value to the customer. This flow creates one part of the value stream, which we can define as broadly as the order of raw material required to deliver the product to the customer's doorstep.

Although the general conceptualization of a manufacturing process may be from receiving dock to shipping dock, the principle of pull helps us to look at the flow beginning from the time a customer requests a product. That product order triggers a demand for the preceding (upstream) process. This is much like a railroad locomotive engine at the beginning of a long line of a railroad car in which the pull from the locomotive engine pulls the attached car, which in turn pulls the next car all the way to the end of the train. Toyota is known for its use of a just–in-time manufacturing system, they look at inventory that is sitting between manufacturing steps as waste. Their rationale for reducing that inventory and working toward just in time is twofold. One reason is that there exists a physical cost as well as the possibility of defective material. The second reason, which is just as critical, is that the inventory basically covers up the waste that might exist on either side of it. Reducing the inventory exposes the issues that can then be addressed as part of lean efforts.

Lean in Development

In a manufacturing process the flow of the value add may be the flow of a physical component, whereas in development the flow of work is the creation of knowledge from learning. This knowledge then results in a solution that delivers value to the customer. This flow of knowledge is how we as engineers do our work. We can look at this in a very specific manner, such as "How much knowledge did I create today or this week?" Then, from a lean perspective, "What slowed down my learning?" and "How can I learn faster?" As we mentioned above, and will show below, the principles of Lean Development can be viewed as a set of "personal productivity" tools. This personal productivity allows us to increase our value add to the business.

Let's take the two general areas, the type of waste and the opportunity to add more value, and apply them to our daily work.

A simple goal of productivity, at the personal level, might be:

- Maximize value add and innovation while "going home on time."

Throughout the book, we have explicitly used the concept of delivering better and faster results as an expectation of utilizing Lean Development methods. If we take the three types of work, Value Add (VA), Non-Value Add but Necessary (NVA but Necessary), and Non-Value Add (NVA), and apply them at the individual level, we would see a flow of our "workday" on the vertical (*y*) axis of Figure A.1.

In Figure A.1, the stacked bars such as NVA, NVA but Necessary, and VA are not stacked randomly. Value Add, in this model, is shown to happen at the end of the day. This is not due to knowing exactly how things happen, because we know the three forms of work are interspersed throughout the day, but because the Value Add to our job is the part that keeps us from going home on time. We need to get it done in order to leave work – or leave our computer. In contrast, if we knew that everything at the end of the day was NVA, we would just go home. However, we don't know that or it did not happen that way so we continue "working" until we are done.

So now, let's look at this from the goal of "Going home on time." The first step is to internalize that we want to focus our effort on eliminating the stuff that doesn't matter or, in the words of lean, eliminate the waste.

Figure A.1 shows the "current state" on the left, listed as "Current Non-Lean." Then, with the progression of time over the course of months, moving to the right on the x-axis, we end up "going home on time" while adding more value (as shown in the far-right bar, the "Target State"). Even at the target state, we show a small bar of NVA because NVA (waste) always exists. We may decide to ignore it or get around to it at some time. Further, NVA but Necessary may be completely out of our control.

The sequence of these steps, like the stacked bars, is not random either. The first step is to reduce Non-Value-Add activities because this step is the easiest and attacks the problem head-on. Your goal is to get your head above water or, in this case, "go

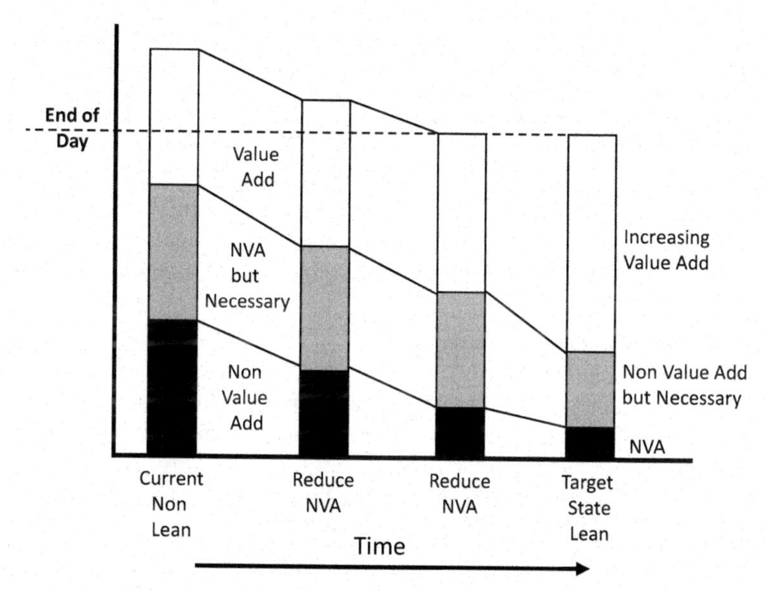

FIGURE A.1 Increasing value add while going home on time.

home on time." The next step is to reduce the Non-Value Add but Necessary time. Both of these actions allow for an increase in Value Add work. Generally, there is minimal return when we think about "improving the efficiency of Value Add work." This is not to say that it can't be done; however, there are typically bigger opportunities that are more easily achieved in the organization, specifically those that involve eliminating other NVA activities.

Beyond working on NVA first, none of this is science. There is no piece of software that will do this. Four elements fundamentally drive it.

1. The acceptance of the situation and the desire to address it.
2. An understanding of the methods of lean or having someone to guide you.
3. The development of insight and critical thinking as to how to look at the problem.
4. The fortitude to work on it and seek help in doing it.

If we now loop back to our causal diagram (Figure 6.2), the overt effort by someone (maybe someone being guided by a technical coach or a manager) begins with the bubble of Work Methods and focuses on creating Satisfying Work. As we eliminate the NVA and make more time for VA, we have the opportunity to create Balanced Work. This "Balanced Work" gives us the "Opportunities" to take on other work that can enable "Control over Career." We can also see that the "Work Methods" that enable us to go home on time would directly feed the "work climate" through, at a minimum, "work–life balance." For these reasons, "Work Methods" was one of the three root causes identified in the causal diagram along with Reusable Knowledge and Role Models. Work Method focuses on two key elements: what the work is and how the work is done.

The following are the principles of Lean Development.

THE SIX PRINCIPLES OF LEAN DEVELOPMENT

In Chapter 11, we provided an overview of Lean Development and its six principles to change how the work is done [8]. For the remainder of this appendix, we will build and expand on the six principles to facilitate a broader understanding of Lean Development. The six principles of Lean Development are listed below:

1. Creation of reusable knowledge
2. Cadence, flow, and pull
3. Visual management (added to Allen Ward's original five principles)
4. Entrepreneurial system designer (a core element of this principle is understanding value)
5. Set-based concurrent engineering (or Set-Based Design, for our use)
6. "Build Teams of responsible experts" [8, p. 1]

Although these principles are generally independent, we have placed them in this order because we believe they build off of each other much like the five principles of lean listed above from the book "Lean Thinking" [7]. Specifically, we view reusable knowledge as the foundation of Lean Development. Unlike a manufacturing process in which you can see the flow, in development there is little to see if it is not written down in a reusable form. From reusable knowledge comes cadence, flow, and pull. With the process established, the team now understands the rules of the game or the sequences of plays to be executed – communicating precisely what starts the work and how it progresses through the process. In the broad sense of its use, visual management stresses the importance that if you can't see the work, you can't manage it. In our case, the first step in visual management is the use of A3 reports. With the principles of knowledge, flow, and visibility, a system designer can then begin to establish the value to be delivered and the needed learning. Throughout the development work effort, we focus on our overarching goal – namely that the execution of actually doing the work enables the success of teams and increases the depth of their knowledge and skills. These teams of experts will ultimately deliver the value to the customer.

Additionally, we propose that in this approach, principles 1, 2, and 3 form an infrastructure to manage the development work. Principles 4, 5, and 6 enable value creation. They are driven by the people actually doing the work and responsible for the process of learning.

PRINCIPLE 1: CREATION OF REUSABLE KNOWLEDGE – THE POWER OF A3 REPORTS

Using an A3 Report is the simplest way to create reusable knowledge. The single side of an A3 or 11 × 17 inch piece of paper forces clarity and conciseness in stating the problem and delineating the work to resolve it. The structure of A3s is engrained in Toyota's corporate culture and follows the problem-solving process of Edwards Deming: Plan–Do–Check–Act (PDCA) [9, 10]. This then becomes a PDCA cycle. In Japan, this is sometimes referred to as the Deming cycle or Shewhart cycle.

The PDCA cycle is managed as a series of repeating steps. For example, when ending the Act step, the problem solver naturally reviews the Plan step and repeats the cycle. In addition, inside any of the individual PDCA steps there are naturally additional PDCA cycles occurring. The A3 sheet is then set up in the form of a PDCA, with the Plan on the left side of the page and the Do–Check–Act on the right side [10, p. 31].

There are many books and templates available describing the process of creating an A3. Some can be used by basically filling in the boxes, others are set up more as free form but still live into the intent of the PDCA structure. A general working example of a sheet may look something like Figure A.2. Sobek's and Smalley's book, *Understanding A3 Thinking – A Critical Component of Toyota's PDCA Management System* (2008), is an excellent book that gives detailed information about the method of an A3. The book additionally includes great templates [10].

The left side of an A3 (Figure A.2) is focused on the situation and the proposal. The right side has the expected flow of learning and outlines the actions. Toward the bottom-right corner of the sheet, the learnings and next steps are communicated. The sheet is a working document. It is a sheet that you may expect to use for days,

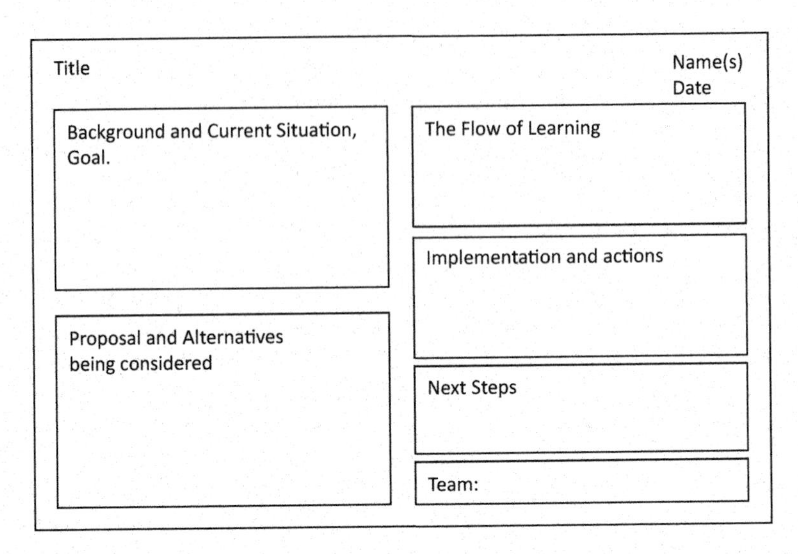

FIGURE A.2 Example of working template of an A3.

weeks, or months, depending on the complexity and size of the problem or task being worked on. We include a box labeled "Team," lower bottom right, because a significant amount of development is done through the efforts of a Team. In addition to recognizing people's contribution, which is important, the individuals listed demonstrate the breadth of knowledge that was incorporated into the sheet. This is also a concrete way of promoting team development through the work.

The key to the successful use of an A3 is to:

1) Start the sheet then
2) Use the sheet
3) Work toward creating a sheet that is worthy of an "A" letter grade. Just like school.

A3s Compared to Other Forms of Knowledge

Although the difference between an 11 × 17 inch or A3 size sheet of paper and a multi-page text document may seem like a formatting choice, the constraint of a single sheet of paper is a powerful tool to force clarity and conciseness. In addition, with all of the information sitting in one view (e.g., not spread across multiple pages), the reader and the problem solver see the entire picture in one view versus spread across many sheets.

Creating an A3 (See Chapter 11 for Four Types of A3s)

a. It is one page of 11 × 17 inch or A3, no page 2 (no front and back).
b. From a template or a blank sheet of paper, fill in the boxes or use a free form approach in a PDCA flow.
c. Provide a clear statement of the problem and goal/target, beginning in the upper left corner.

d. Choosing pictures/graphs over words and having a 50/50 split of words to graphics is reasonable.
e. Use active words over passive words.
f. Use fact over opinion (however, opinions can be a hypothesis to test).
g. Nothing smaller than 10 pt font.
h. Use bullets over paragraphs (If you have written a paragraph, just add in bullets between sentences).
i. Put items of focus in outlined boxes.
j. Have clear actions and countermeasures.
k. It should be completely readable when printed as a 11 × 17 inch or A3 size.

For the use of A3s in the day-to-day work environment, the bar for it to be useful is established by those that are directly using it. This means that the author, management team, and peers have direct responsibility for the success of the work and are those that enable its success. A3s are a great place to capture trade-off curves as part of Set-Based Design. The detail data, for those curves, can then be found on a supporting spreadsheet as another form of reusable knowledge.

Ward and Sobek write,

> But, for the knowledge to actually get applied to future projects, it must be actionable. Sifting through large volumes of A3 reports or other technical documents archived on a shared drive to find the few nuggets of knowledge is not readily actionable! Lean managers help their teams consolidate their learning into design guidelines or development checklists
>
> **[11, p. 234]**

To support future use of the knowledge, lean managers contribute to this process by looking at A3s for areas that new learning occurred and then helping the engineer put it in a generic form. This would include removing content that, although critical during development, would not be relevant to someone in the future. This promotes clarity and conciseness. Finally, the lean manager's responsibility is to ensure that the engineer is given time and support to move this newly created knowledge into the organization's knowledge repository.

PRINCIPLE 2: CADENCE, PULL, AND FLOW

This principle targets how the development work happens through time. It focuses on how it starts and what it delivers. This principle can be applied to the physical development of a piece of hardware that will be delivered to the customer or to the learning necessary to make a decision about a business or technical choice.

Development, The Flow of Work

We will break the work flow into six key elements.

a. *A pull event by the business*: This might be a targeted product introduction date, a new business strategy, or addressing a new issue.

b. *Some understanding of the objectives,* development strategies, or basic solution architecture that enable a general view of the development work to be thought about.
c. *Identification of natural integration events* that "pull" together key pieces of the development work and required learning. These are ideally physical in nature (e.g., models, trade-off curves, customer data) and force technical or business reviews.
d. *Identification of knowledge gaps* through an open and honest technical or business conversation about the work that needs to be done and a plan for obtaining needed knowledge. This element would drive the identification of "Sets to be Investigated" for knowledge gap areas.
e. *The identification of risk,* although knowledge gaps (customer or technical), more than likely create risks due to the unknowns, other risks could be targeted as organizational processes. This might include organizational alignment, resource availability, or variability in a specific process.
f. *The identification of development areas* that need detailed development plans (e.g., A3s). This is the place where the thought process of creating the cascading of A3s, Figure 13.1, begins.

Creating an Integration Plan as Part of an A3

Figure A.3 shows the integration plan for a hypothetical product [12]. This plan would be a key element contained within an A3 focused on the development work and integration activities. To the left of the integration plan on the A3, the author would outline key elements such as alignment to the business, the customer need, and objectives. The integration work begins at the right edge with the identification of the pull event which, given something like a consumer product and the holiday selling season, would be driven by the on-shelf date or on-web date. The plan then works upstream (to the left) to identify the major milestones, major work areas, decisions,

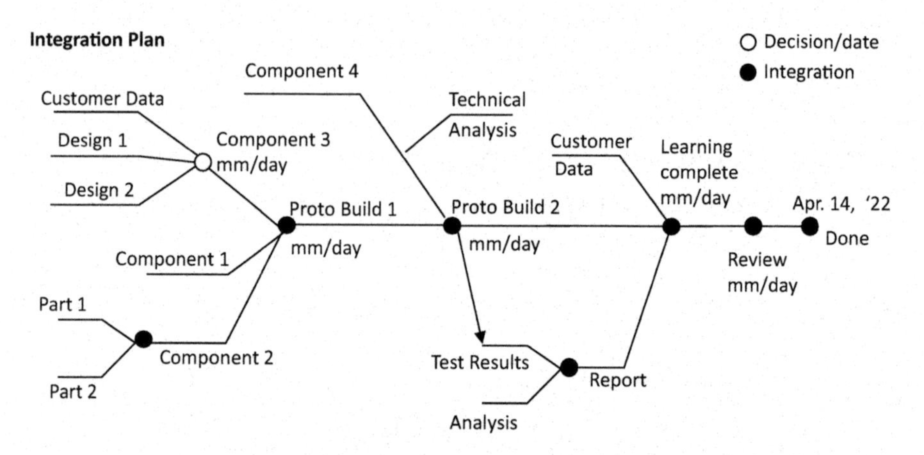

FIGURE A.3 Integration plan contained within an A3.

and how they interrelate. In addition, the plan identifies the integration events that pull specific parts of the development work together such as critical components or customer data with targeted dates. These integration events focus on learning and driving decisions. In addition to technical performance, customer data may be a driving force at any of these decision points. The integration events are focused on build–measure–learn. If these events feel like a "check off" item or the reviewers lack sufficiently deep technical or business knowledge to ask tough questions, the event is not living into the expectation of a technical or business review.

The purpose of this type of diagram is to establish a common understanding of the flow of the work and its critical elements. For complicated development, it should be expected that multiple A3s are created in order to manage specific work identified on this integration sheet. As an example, the author should expect that an open-circle question utilizing Set-Based Design would have a dedicated A3 to manage that work.

A Gantt Chart is not integration plan. A Gantt Chart is a good tool for work that has well-understood tasks, with defined schedules for each element, clear critical path items, limited knowledge gaps, and understood dependencies – little of that exists within development work. A Gantt Chart is a great tool if you are constructing a house but not if you are working to create innovation.

A final thought. Development schedules don't slip past their planned delivery date near the end of the work. They slip throughout the work when, during traditional development work, no one is looking at the work from the perspective of what needs to be learned and by when in order to make decisions to support the schedule. A well-constructed integration plan that incorporates the flow and pull of the work will highlight the areas of highest knowledge gaps in advance of the need. This enables the team to work through those areas on the required schedule or find another way to gain the knowledge. A technically driven integration plan, driven by business and customer need, coupled with Set-Based Design (to be covered below) is a powerful learning method.

PRINCIPLE 3: VISUAL MANAGEMENT

Within a manufacturing process, the visual aspects of the workflow almost come for free. However, in product development, not much may be understood by the organization as a whole, let alone an individual outside the system when the vast majority of the knowledge is sitting inside computers or even on shared drives.

As mentioned in Chapter 11, the challenge with visual management within product development is that work done in this area may seem like extra work without a clear benefit seen at the engineering level. However, that does not make it less important. A prominent example of visual management is the use of an Obeya (big room). The use of a big room can enable the creation of an overall management system [13, p. 95]. However, deciding to implement an Obeya as a part of the initial change management effort to move to Lean Development may be a big and challenging step without having the other principles firmly in place. For example, a weekly standup meeting using an Obeya or Visual Management Board filled with charts and lists that seems to make the management team feel better about the work but does not actually help the development team move faster should be questioned. The key is, how is this work improving the learning rate of the development team.

So, with respect to Visual Management, where does that leave us? Specifically, we fall back to our basic change management approach: start small and ensure that any efforts toward Lean Development benefit the individuals doing the work. The second part of the effort is to take credit for the progress toward this principle that we are already doing, specifically we are going to approach the use of A3s and their review process, and the creation of Trade-Off Curves as examples of Visual Management. The third step is to use these examples of visual knowledge as parts of expanding efforts toward visual management. Explicitly, for an individual work team that sees the need to create a visual management board, maybe in an aisle way, to help provide visibility to the work, start with the work that has already been created. This might be hanging high-level program A3s on the board (wall). To help with readability, we might print those A3s in a larger format (e.g., 17 × 22 inch or A2). This might be a simple way to help partner groups better understand the flow of work and where their work fits in. Then, when partners stop by to ask about the work, it is a quick few steps to walk out into the aisle way to talk through the work directly from the large format printed A3s hanging on the wall. However, the key is to make sure that these versions are kept up to date. This may also be a place to manage the list of tasks to be completed and facilitate a daily standup meeting.

PRINCIPLE 4: ENTREPRENEUR SYSTEM DESIGNER

The (entrepreneur) system designer is just that: she focuses at the system level as part of her overall job. This includes all of the elements that add value for the customer and drive business-level decisions that a person is responsible for or influences. It also includes all aspects of the development work that a person is responsible for.

For this system designer, in the auto industry, Toyota developed the role of a Chief Engineer (CE) for a new car model – which has been replicated across numerous industries and firms. "The CE does all the system design activity – designing the vehicle architecture, creating the product plan, deciding vehicle targets and determining the project schedule. In essence he is the master designer" [6, p. 146].

Although the concept of a Chief Engineer is interesting, for our purposes with respect to changing the way engineering is done, we focus this principle on individuals thinking and acting as system designers.

The Individual as a System Designer

Throughout our hypothetical stories in which the engineer was using an A3 to do the work, she was acting as a system designer. In that role, she ensured that the work covered the broad system, was visible, and technically clear to those involved in the work. This began with taking responsibility for sharing her work with peers and managers. Specifically, gathering input into the work and then using the A3 Report, especially the integration map, to manage the work. Through this, she became a capable designer.

THE SYSTEM DESIGNER – INTEGRATING ACROSS DEVELOPMENT ACTIVITIES

For most of this topic, we have spoken of how a system designer needs to look at her work relative to her own personal responsibility or that of the team she leads. Now, we will look at how to approach system design across multiple development activities, specifically if you are a team leader or have been asked to play this role.

This work basically leverages the cascading use of A3s (Figure 13.1), but is intended to align the work of multiple system designers into one cohesive set of development activities that build off each other and enable the success of each other. In the world of development, the use of cohesive development activities intentionally creates powerful teams.

So, how to start? Integrate the design or processes pieces together such that the entire team sees the whole picture. In Figure A.4, the entire system is represented by the A3 labeled "Program A3" which the team created together. This sheet communicates the overall objectives, system design, integration activities, and so on. In addition, the sheet specifically identifies the key development areas. In this case, there are five. Then, through the work of individuals, each development area has an A3 that the Program A3 drives. This is basically the step of cascading A3s as shown in Figure 13.1. In the example shown in Figure A.4, Development Area 1 and Development Area 2 may have very similar work, but very different from Development Area 5. However, all of them support the Program A3.

The development area A3s may have a common look and feel, but you would not expect them to look identical. The system designer then facilitates a structured review of all the A3 sheets for Development Areas 1 through 5. Through this review process, labeled "Structured Review Process by Team," the owners of the A3s learn from each other and provide critical feedback. They are a team, not just in name, but in action and acceptance of a focus on the overall goal.

Concrete actions through how the work is done, such as the structured review of the team's development sheets, are a tangible work method. These actions then enable one of our later principles, principle 6, building teams of responsible experts.

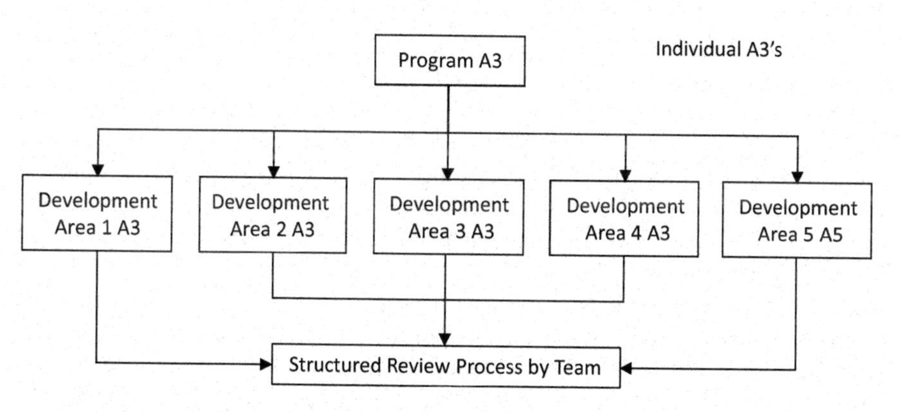

FIGURE A.4 Technical review across development areas.

PRINCIPLE 5: SET-BASED DESIGN (INCLUDING TRADE-OFF CURVES)

Figure 11.1 shows a representative model of the flow of learning for Set-Based Design. This is in direct conflict with traditional development shown in Figure 9.1, which uses the method Build–Test–Fix. In Chapter 9, we have described how elementary-age students are being taught Build–Test–Fix, but there is no question that Set-Based Design is harder. Set-Based Design requires a higher level of rigor and a higher level of thinking. It requires the designer to think about and understand the design space and to set aside their bias about the "best option." The designer has to understand how each of the choices in the set relates to each other and the expected performance. Set Base Design requires discipline to trust that as you work and evaluate each option, you are learning about what makes the best design.

The starting point for this principle is to put it on an A3, structured as above. Then, create some form of graphical representation of the options being considered or the design space that can help map out the flow of the learning. Figure A.3 is one example of a graphical representation, the integration plan could show the learning steps for each option. At each step of learning, the performance criteria can then be established. The learning should begin with the highest level of risk, the largest knowledge gap, or the lowest level of confidence. For each option identified, a simple table that lists out the required learning and results can be used to guide the engineering team through the learning steps. The table may have rows of performance attributes while utilizing the columns for design choices. Then, within the table, a simple color coding of green, yellow, and red may highlight the performance result for a specific design choice for each attribute. The purpose of this step is to be able to look at all the options together as the learning occurs. Additionally, it may very well be that the option with the highest risk, the largest knowledge gap, and the lowest level of confidence rises to the top of greatest customer value because of the learning conducted at the beginning of the work.

TRADE-OFF CURVES

As we add a deeper level of knowledge from each of the learning cycles, there should be an emphasis on constructing Trade-Off Curves. Using Figure 11.2, we could imagine that this example is a comparison of two designs and a measurement of their performance for a range of inputs. Designs A and B may differ in cost, size, or rated performance life. This graph allows the developers to work through the process of how to maximize the Value Add, which could be creating Design C that sits right in-between Designs A and B, by adding a new curve on the graph and delivering a higher level of customer value.

Additionally, if an engineer is given a trade-off curve of just one design and asked to improve its performance by 20% (which means shift the curve to the right or left), the engineer already has half the information. Basically, it is like having the answers to all the even-number problems in the back of a textbook while being assigned the odd-number problems for homework.

PRINCIPLE 6: TEAMS OF RESPONSIBLE EXPERTS

As covered in Chapter 11, in Allen Ward's and Durward Sobek's book *Lean Product and Process Development*, they list the four tasks of a responsible expert as: "Focus on overall project success, create new knowledge, communicate it, and represent it (conflict to consensus)" [11, p. 227]. Toyota's people development methods and focus is highly researched. Sobek characterized the Toyota engineer as "technical experts with a system-level understanding of how they and their parts fit into the bigger picture" [6, p. 25].

In addition to a firm's leadership team developing a passion for creating teams of responsible experts, the vision of the engineering system in this area is one in which the design and execution of it create teams of responsible experts through the growth of individuals. In other words, although there are specific actions targeted at the growth of individuals and the development of teams, the majority of progress toward this principle comes from executing the system. For example, a system designer looking across multiple development areas as part of the structured review promotes the creation of responsible experts. A secondary, but equally important, aspect of this type of structured work (from Figure A.4) is that the team members can transfer work back and forth among themselves if the work situations require it. They have clarity and understanding of each other's work through the technical review and the creation of A3s,

In Chapter 14, and our story "Learning by Reading" which describes Control over Career due to the availability of reusable knowledge, we further see how the design of the system enables this principle. In Chapter 15, we further expanded this story to overtly bring in the female technical lead to create role models. Both of these stories demonstrate how the implementation of principles of Lean Development can help to create responsible teams.

The Learning Rate of a Team

In Chapter 10, we wrote of how the increased learning rate by an individual through the methods of Lean Development translates into the business seeing customer solutions sooner (Exercise 10.1). In this exercise, when people are working on the 45-degree line, they achieve 100% of the required knowledge ~40% sooner than someone working on the 30-degree line. This increased learning rate comes from the use of the specific methods within Lean Development that we covered above and is driven by the elimination of waste within the system – the waste that slows down learning. When someone finishes the work sooner, they are able to move onto other learning opportunities (Exercise 10.2) and, as a result, see ~70% additional learning compared to someone working on the 30-degree line during the same time period.

The ability to learn faster and have additional learning opportunities facilitates the development of teams of responsible experts. Figure A.5 shows how Exercise 10.1 and Exercise 10.2 integrate together. The nuance of Figure A.5 is that anyone who has been on the 45-degree line for multiple years has a knowledge depth that is significantly deeper than someone on the 30-degree line. When a product program is

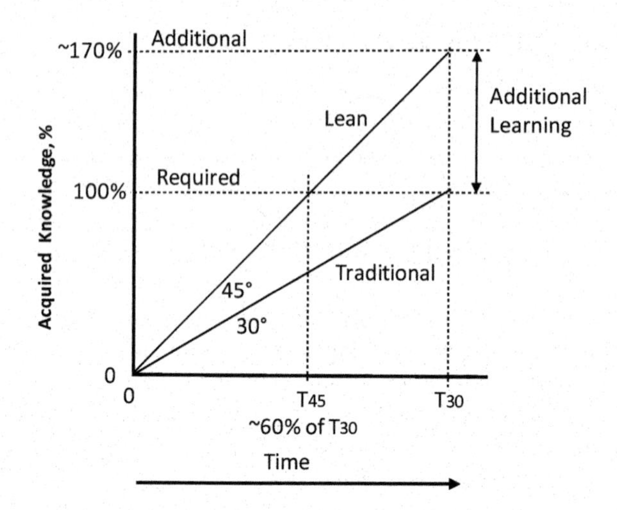

FIGURE A.5 Knowledge acquired verses time.

started with 45-degree individuals, that team has a head start, moves faster to completion, and requires less resources in doing the work because each individual has a higher level of knowledge. This gets us back to the 1987 research which showed "that Japanese projects were completed in two-thirds the time and with one-third the engineering hours of the non- Japanese projects" [1, p. 740]. We began with this quote in the section "The Business Case for Lean Development," as the driver for an organization's investment in Lean Development and we end with this quote within this section as to what delivers these types of results, a team of responsible experts.

SUMMARY

There are numerous books that describe Lean Development in significant detail. Many of them weave the various attributes of the methods together. Our belief, however, is that being overt about focusing on each principle allows an individual to break down the overall use of Lean Development into individual pieces. This allows for deliberate coaching. This approach may be more akin to how some engineering concepts are learned, understand the principles, and then apply them to real-world problems. Additionally, this gives a leader the opportunity to basically give a letter grade for the execution of each principle. In the beginning, no one should expect an A for each of these principles and each principle should be graded against reasonable targets, but progress against each of them should be expected.

WORKS CITED

1. K. B. Clark, T. Fujimoto, and W. B. Chew, "Product Development in the World Auto Industry," Brookings Papers on Economic Activity, 1987.

2. K. B. Clark, and T. Fujimoto, *Product Development Performances – Strategies, Organization, and Management in the World Auto Industry.* Boston, MA: Harvard Business School Press, 1991.
3. J. P. Womack, D. T. Jones, and D. Roos, *The Machine that Changed the World.* New York: Rawson Associates, 1990.
4. D. K. Sobek II, J. Liker, and A. C. Ward, "Another Look at How Toyota Integrates Product Development," *Harvard Business Review*, July–August 1998.
5. D. K. Sobek II, A. C. Ward, and J. K. Liker, "Toyota's Principles of Set-Based Concurrent Engineering," *Sloan Management Review – MIT*, vol. 40, no. 2, pp. 67–83, Winter 1999.
6. D. K. Sobek II, *Principles that Shape Product Development Systems: A Toyota – Chrysler Compare.* Ann Arbor, MI: UMI, 1997.
7. J. P. Womack, and D. T. Jones, *Lean Thinking – Banish Waste and Create Wealth in Your Corporation.* New York: Simon & Schuster, 1996.
8. A. C. Ward, *The Lean Development Skills Book.* Ann Arbor, MI: Ward Synthesis, Inc, 2002.
9. W. E. Deming, *Out of the Crisis.* Cambridge: Massachusetts Institute of Technology, 1982, pp. 88.
10. D. K. Sobek II, and A. Smalley, *Understanding A3 Thinking.* Boca Raton, FL: Productivity Press, 2008, pp. 31.
11. A. C. Ward, and D. K. Sobek II, *Lean Product and Process Development.* Cambridge: Lean Enterprise Institute, 2014.
12. The basic format of the integration plan was created by Mark Miranda of Vancouver, WA, Hewlett-Packard, in support of a program Bob was leading in 2006.
13. J. M. Morgan, and J. K. Liker, *Designing the Future – How Ford, Toyota, and Other World-Class Organization Use Lean Product Development to Drive Innovation and Transform Their Business.* New York: McGraw Hill Education, 2019.

Appendix B – Bringing Lean Development to a University Engineering Program

A Consideration for Teaching Lean Development and A3 Reports

Objective: Add Lean Development and the tool of A3 reports to a university engineering curriculum program. The benefits are threefold:

1. Teaching the skills and work methods of doing the work as an engineer better and faster.
2. Delivering higher value and more innovation to the organization.
3. Contributing to an organization's moves toward Lean Development or possibly being a catalyst for starting it.

   (Software students are learning the methods of Agile and Scrum, engineering students should be learning Lean Development.)

As covered in Chapter 10, while Lean Production or Manufacturing is focused on the application of lean to a process within a factory or some other system that delivers customer value, Lean Development is focused on how an engineer does the work and the system of engineering that they do that work within. In a very simplistic example, teaching Lean Development to students is no different from teaching the latest computer-aided design (CAD) tool.

When we think about adding something new to an educational program that is already full, we need to start with the perspective that the addition of this knowledge is not competing with the most important content that is covered, it is competing with what is the least important – given the current situation. Although this sounds harsh, engineering programs have dealt with this for decades. As new automation, technology, tools, and knowledge are available, it naturally replaces areas that are less important or obsolete.

The methods and benefits of Lean Development have been researched and written about for decades with dozens of books written on the topic.

Suggested approach:

1. As part of a Freshman Introduction to Engineering class:
   a. Teach the method of the A3 problem-solving process following the process of Plan–Do–Check–Act or some other basic problem-solving approach.
   b. Students create their first A3 report.
      i. Each student or a small team would create an A3 report to solve a problem or communicate a situation that is relevant to the class or the program they are part of.
      ii. The creation of these sheets facilitates the coaching process.
      iii. The instructor would conduct a review of the sheet, returns it marked up to the student to be corrected, and then there is a final review for a letter grade.
      iv. Consider skipping any work in creating a slide deck for presenting content and have them present from their A3.
   c. Teach the method of causal diagrams: ideally incorporating the method into the creation of the A3 for the class. This could be expanded to the creation of trade-off curves.
2. For each subsequent class term, identify one class within a term in which the student would create an A3 as part of that class. By the time a student is entering his or her senior year, they would have created six or more A3 reports.
3. For individual classes that have a design component to them, establish what Lean Development principle should be brought into the class.
   a. This may be the point to highlight the principle of Lean Development and the specific principle to be worked on.
   b. As a part of a class prior to senior year, teach the method of trade-off curves and link them back to how a causal diagram helps to find them.
   c. For programs in which the method of trade-off curves are already taught, great.
4. As part of a senior design project:
   a. Teach the principles of Lean Development. The key principles that students would utilize are:
      i. Creation of a A3 development proposal. The instructor would review these as part of the development process.
      ii. Mapping out the development flow and most importantly the steps of identifying knowledge gaps.
      iii. The use of Set-Based Design and the creation of Trade-Off Curves. (This should be directly compared to the traditional approach of Build–Test–Fix.)
      iv. The importance of being a System Designer.
      v. The focus on building a team of responsible experts.

b. Expect the use of A3 reports as part of the development work. Establish key milestones that A3 reports would be reviewed. Ask the question "Does that area of development need an A3?"
c. Near the end of the project, move the development work captured on A3 reports to an A3 that clearly capture the knowledge that was learned in a reusable form.
d. Utilize the presentation of key A3 reports as part of the final presentation – consider forgoing any use of a slide deck presentation.

The end result:

1. Each graduating student will be well-versed in the creation of A3 reports and how they can be used in their future employment.
2. A student will understand the principles of Lean Development and specifically: reusable knowledge, set-based design along with trade-off curves, and the expectation of being a system designer.
3. As part of their senior project, they will understand how the use of Lean Development drove that work.

A final thought,
For the one in five students who are women in the program, they will have a firsthand understanding of how to do the work differently from the traditional methods and, at a minimum, know what they might want from their first job.

Appendix C – Books to Consider

WOMEN IN TECHNOLOGY

There are a multitude of good books on the market that can be beneficial to women in technology, here is a list to consider.

Bush, Pamela McCauley. *Transforming Your STEM Career Through Leadership and Innovation – Inspiration and Strategies for Women.* 2013, Academic Press.

Fletcher, Joyce K. *Disappearing Acts – Gender, Power, and Relational Practice at Work.* 1999, Massachusetts of Technology.

Fournier, Camille. *The Manger's Path – A Guide for Tech Leaders Navigating Growth & Change.* 2017, O'Reilly Media.

Pritchard, Peggy A. and Christine S. Grant. (editors). *Success Strategies from Women in STEM – A Portable Mentor.* 2015, Academic Press.

Slocum, Stephanie. *She Engineers – Outsmart Bias, Unlock Your Potential, and Create the Engineering Career of Your Dreams.* 2018.

Williams, Joan C. and Rachel Dempsey. *What Works for Women at Work – Four Patterns Working Women Need to Know.* 2014 and 2018, New York University Press.

Williams, F. Mary and Carolyn J. Emerson. *Becoming Leaders – A Practical Handbook for Women in Engineering, Science, and Technology.* 2019, 2nd edition, American Society of Civil Engineers.

LEAN DEVELOPMENT

There are dozens of books on the subject of lean development, here is a list to consider.

Cloft, Penny W. and Michael N. Kennedy and Brian M Kennedy. *Success Is Assured – Satisfy Your Customers on Time and on Budget by Optimizing Decisions Collaboratively Usin g Reusable Visual Models.* 2019, Routledge.

Kennedy, Michael N. *Product Development for the Lean Enterprise – Why Toyota's System Is Four Times More Productive and How You Can Implement It.* 2003, The Oaklea Press.

Mascitelli, Ronald. *Mastering Lean Product Development – A Practical, Event-Driven Process for Maximizing Speed, Profits, and Quality.* 2011, Technology Perspectives.

Majerus, Nobert. *Lean-Driven Innovation – Powering Product Development at the Goodyear Tire & Rubber Company.* 2016, CRC Press.

Morgan, James M. and Jeffrey K. Liker. *The Toyota Product Development System – Integrating People, Process, and Technology.* 2006, Productivity Press.

Morgan, James M. and Jeffrey K. Liker. *Designing the Future – How Ford Toyota, and Other World-Class Organization Use Lean Product Development to Drive Innovation and Transform Their Business.* 2019, McGraw Hill Education.

Oosterwal, Dantar P. *The Lean Machine – How Harley-Davidson Drove Top-Line Growth and Profitability with Revolutionary Lean Product Development.* 2010, AMACOM.

Radeka, Katherine. *The Mastery of Innovation – A Field Guide to Lean Product Development.* 2013, CRC Press.

Sobek II, Durward K. and Art Smalley. *Understanding A3 Thinking – A Critical Component of Toyota's PDCA Management System.* 2008, CRC Press.

Ward, Allen C. and Durward K. Sobek II. *Lean Product and Process Development.* 2014, 2nd edition, Lean Enterprise Institute.

Ward, Allen C. (with contributions by Dantar P. Oosterwal and Duward K. Sobek II). *Visible Knowledge for Flawless Design – The Secret behind Lean Product Development.* 2018, Taylor and Francis Group.

Author Bios

ROBERT N. STAVIG, BRUSH PRAIRIE, WA

During his 35-year career with Hewlett-Packard, Bob spent the last 20 years in Product Research and Development in various management and technical roles. During 16 years of that time, he utilized the principles and methods of Lean Development in the delivery of nearly a dozen areas of work within product programs, process development, and technical asset development. Bob considers himself an engineer first and a manager second. In addition, Bob has 15 years of experience in the areas of manufacturing development, factory support, factory operational management, and worldwide manufacturing leadership. He has over 20 years of experience working with international partners. Bob has a degree in Mechanical Engineering from Washington State University and is a Certified Six Sigma Black Belt by the American Society of Quality. Bob has a certification in Lean Product Development from the University of Michigan and is a Certified Scrum Master. He has taught Lean Development methods in multiple venues including Lean Product and Process Development Exchange (LPPDE) Virtual Summits. As of January 2022, Bob is now serving on the board of LPPDE.

ALISSA R. STAVIG, MD, BILLINGS, MT

Alissa graduated from Washington University in St. Louis with an A.B. in anthropology and physics and received her Doctor of Medicine from the Duke University School of Medicine in 2017. She graduated in June 2022 from a combined five-year residency program in Internal Medicine and Psychiatry at the Duke University School of Medicine in Durham, NC. Alissa's training includes clinical experiences in inpatient and outpatient settings in both disciplines and educational experiences in advocacy, interdisciplinary and integrated healthcare, quality improvement, and medical education. During her residency training, she experienced the benefit of developing clinical and leadership skills through a system with graduated responsibility which included leading clinical teams of medical students and first-year residents. Alissa's background in anthropology and physics in combination with her clinical experience throughout medical school and residency have shaped her passion for understanding the individual within their broader social, historical, and cultural contexts.

Contributor List

Debra Blakewood has degrees in Mechanical Engineering and Engineering Management and is a certified PMP (Project Management Professional). Following 20 years of engineering at Hewlett-Packard, she now applies problem-solving approaches and project management to software development.

Rose Miranda Elley has a BS in Electrical Engineering, an MBA, an MA in Leadership, and is a certified PMP (Project Management Professional). Her 20-year engineering career spans Hewlett-Packard and Intel. She is currently caring for her deeply loved mom with Alzheimer's.

Jessica L. Wanke has nine years of elementary education teaching experience with bachelor's degrees in Elementary Education and Spanish, and a Master's in English Language Learning. She has a four-year-old son and a two-year-old daughter.

Index

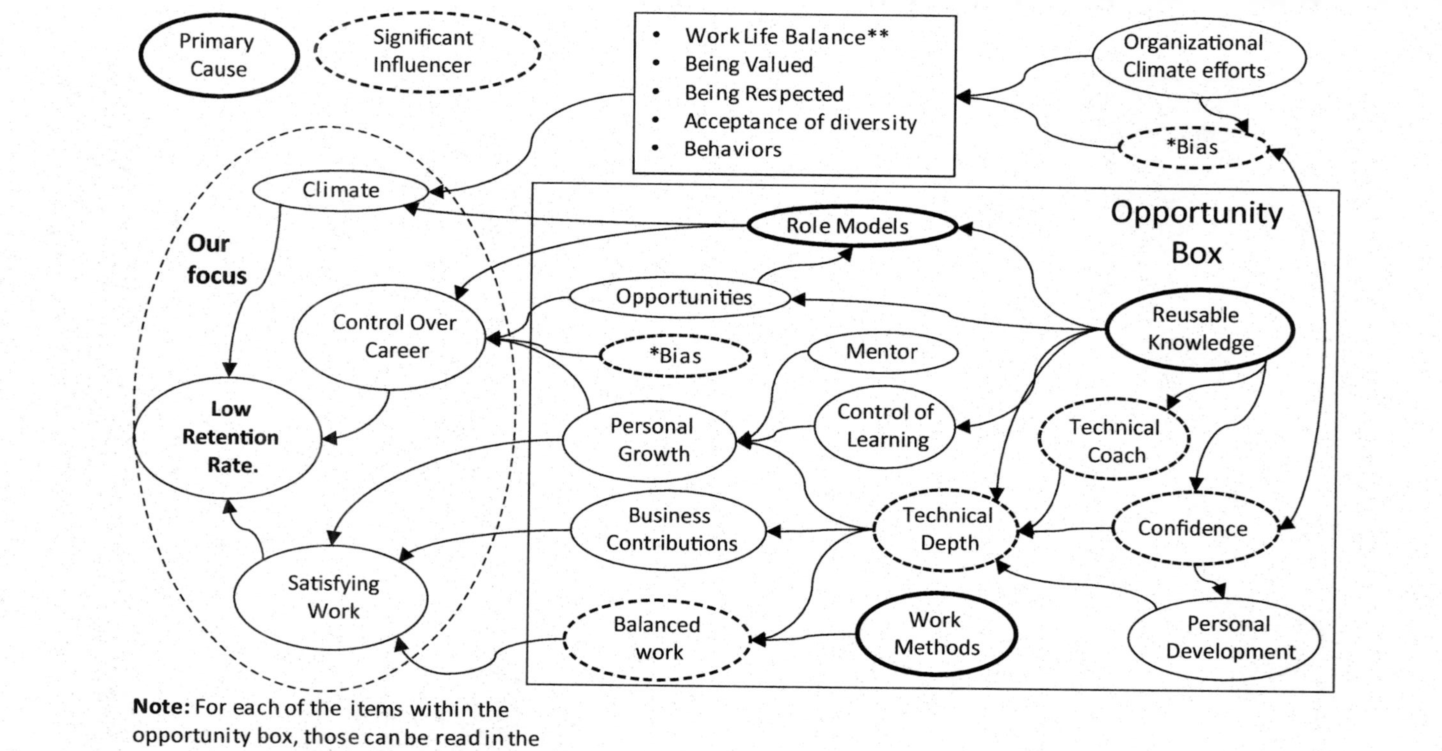

FIGURE 6.2 Causal diagram for Low Retention Rate (*bias can occur everywhere. ** We will include facilitating part-time work in work-life balance).